APPLIED VIROLOGY RESEARCH

Volume 3

New Diagnostic Procedures

APPLIED VIROLOGY RESEARCH

Editor-in-Chief: *Edouard Kurstak, University of Montreal, Montreal, Quebec, Canada*
Series Editors: *R. G. Marusyk, University of Alberta, Edmonton, Alberta, Canada*
F. A. Murphy, University of California, Davis, California
M. H. V. Van Regenmortel, Institute of Molecular and Cellular Biology, Strasbourg, France

Volume 1 NEW VACCINES AND CHEMOTHERAPY
Edited by Edouard Kurstak, R. G. Marusyk, F. A. Murphy, and M. H. V. Van Regenmortel

Volume 2 VIRUS VARIABILITY, EPIDEMIOLOGY, AND CONTROL
Edited by Edouard Kurstak, R. G. Marusyk, F. A. Murphy, and M. H. V. Van Regenmortel

Volume 3 NEW DIAGNOSTIC PROCEDURES
Edited by Edouard Kurstak, R. G. Marusyk, F. A. Murphy, and M. H. V. Van Regenmortel

A Continuation Order Plan is available for this series. A continuation order will bring delivery of each new volume immediately upon publication. Volumes are billed only upon actual shipment. For further information please contact the publisher.

APPLIED VIROLOGY RESEARCH

Volume 3

New Diagnostic Procedures

Edited by

Edouard Kurstak

University of Montreal
Montreal, Quebec, Canada

R. G. Marusyk

University of Alberta
Edmonton, Alberta, Canada

F. A. Murphy

University of California
Davis, California

and

M. H. V. Van Regenmortel

Institute of Molecular and Cellular Biology
Strasbourg, France

PLENUM MEDICAL BOOK COMPANY
NEW YORK AND LONDON

Library of Congress Cataloging-in-Publication Data

New diagnostic procedures / edited by Edouard Kurstak ... [et al.].
p. cm. -- (Applied virology research ; v. 3)
Includes bibliographical references and index.
ISBN 0-306-44668-5
1. Virus diseases--Molecular diagnosis. I. Kurstak, Edouard.
II. Series.
[DNLM: 1. Virus Diseases--diagnosis. W1 AP516R v.3 1994 / WC 500
N5324 1994]
RC114.5.N49 1994
616.9'25075--dc20
DNLM/DLC
for Library of Congress 94-26646
CIP

ISBN 0-306-44668-5

233 Spring Street, New York, N.Y. 10013

Plenum Medical Book Company is an imprint of Plenum Publishing Corporation

Printed in the United States of America

Contributors

Pertti Arstila • Department of Virology, University of Turku, Kiinamyllynkatu 13, SF–20520 Turku, Finland

Robert D. O. Devine • Provincial Laboratory of Public Health, University of Alberta, Edmonton, Alberta, Canada T6G 2J2

Lars G. Fägerstam • Pharmacia Biosensor AB, S–751 82 Uppsala, Sweden

Thomas R. Gingeras • Baxter Diagnostics, Inc., Life Sciences Research Laboratory, P. O. Box 91042, San Diego, California, U. S. A. 92191–0492

M. Hayami • Institute for Virus Research, Kyoto University, 606 Kyoto, Japan

Pirkko Heino • Karolinska Institute, Stockholm, Sweden

Walter Hinderer • Biotest AG, Research Department, Landsteinerstr. 5, Dreieich, Germany

A. Hossain • Department of Microbiology and Immunology, Faculty of Medicine, University of Montreal, Montreal, Quebec, Canada

M–L. Huhtala • Labsystems Research Laboratories Oy, SF–0080, Helsinki, Finland

V. Hukkanen • Department of Virology, University of Turku, Turku, Finland

X. Imai • Institute for Virus Research, Kyoto University, 606 Kyoto, Japan

S. Kontio • Labsystems Research Laboratories Oy, SF–0080, Helsinki, Finland

M. Korkolainen • Labsystems Research Laboratories Oy, SF–0080, Helsinki, Finland

Christine Kurstak • Department of Microbiology and Immunology, Faculty of Medicine, University of Montreal, Montreal, Quebec, Canada

Edouard Kurstak • Department of Microbiology and Immunology, Faculty of Medicine, University of Montreal, Montreal, Quebec, Canada

Deborah Y. Kwoh • Baxter Diagnostics, Inc., Life Sciences Research Laboratory, P. O. Box 91042, San Diego, California, U. S. A. 92191–0492

A. Närvänen • Labsystems Research Laboratories Oy, SF–0080, Helsinki, Finland

Gerald Schochetman • Division of HIV/AIDS, Centers for Disease Control, United States Public Health Service, Department of Health and Human Services, Atlanta, Georgia

Günter Siegl • Institute for Clinical Microbiology and Immunology, Frohbergstr. 3, St. Gallen, Switzerland

John J. Sninsky • Department of Infectious Diseases, PCR Division, Cetus Corporation, Emeryville, California

Danièle Van Beers • Hôpital Universitaire Brugman, Université Libre de Bruxelles, Bruxelles, Belgium

Christian Vandenvelde • Hôpital Universitaire Brugman, Université Libre de Bruxelles, Bruxelles, Belgium

L. Wray • Probe Diagnostics, Abbott Laboratories, Abbott Park, Illinois, U. S. A. 60064–3500

Preface to the Series

Viral diseases contribute significantly to human morbidity and mortality and cause severe economic losses by affecting livestock and crops in all countries. Even with the preventive measures taken in the United States, losses caused by viral diseases annually exceed billions of dollars. Five million people worldwide die every year from acute gastroenteritis, mainly of rotavirus origin, and more than one million children die annually from measles. In addition, rabies and viral hepatitis continue to be diseases of major public health concern in many countries of the Third World, where nearly 300 million people are chronically infected with hepatitis B virus. The recent discovery of acquired immunodeficiency syndrome (AIDS), which is caused by a retrovirus, mobilized health services and enormous resources. This virus infection and its epidemic development clearly demonstrate the importance of applied virology research and the limits of our understanding of molecular mechanisms of viral pathogenicity and immunogenicity.

The limitations of our knowledge and understanding of viral diseases extend to the production of safe and reliable vaccines, particularly for genetically unstable viruses, and to antiviral chemotherapy. The number of antiviral drugs currently available is still rather limited, despite extensive research efforts. The main problem is finding compounds that selectively inhibit virus replication without producing toxic effects on cells. Indeed, the experimental efficacy of several drugs, for example, new nucleoside derivatives, some of which are analogues of acyclovir, makes it clear that antiviral chemotherapy must come of age because many new compounds show promise as antiviral agents. In the field of antiviral vaccine production, molecular biologists are using a wide variety of new techniques and tools, such as genetic engineering technology, to refine our understanding of molecular pathogenicity of viruses and of genetic sequences responsible for virulence. Identification of genes that induce virulence is vital to the construction of improved antiviral vaccines.

Novel types of vaccines are presently receiving particular attention. For example, the protein that carries the protective epitopes of hepatitis B virus, which is produced by expressing the appropriate viral gene in yeast or in mammalian cell systems, is now available.

Another group of new vaccines are produced by using viruses as vectors for the expression of genes. Vaccines for rabies, influenza, respiratory syncytial disease, hepatitis B, herpes infection, and AIDS, which are based on greatly enhanced expression of the viral genes in vectors are being tested. Also of interest is baculovirus, an insect cell vector system now used in the development of recombinant DNA vaccines for a variety of important human and animal virus diseases. This system yields very large quantities of properly processed and folded proteins from the rabies, hepatitis B, AIDS, and Epstein–Barr viruses, among others.

The synthetic peptides, which act as specific immunogens, have also received attention as new antiviral vaccines. The recent experimental performance of new synthetic peptides of foot–and–mouth disease virus, as well as peptide–based vaccines for poliovirus, rotavirus, hepatitis B, and Venezuelan equine encephalitis virus, gives strong support for this group of specific immunogens. However, testing of these synthetic peptide vaccines is in the early stages and future research will have to answer several questions about their safety, efficacy, and immune responses.

Current attempts at developing synthetic vaccines are based either on recombinant DNA technology or on chemical–peptide synthesis. Several virus proteins have been produced in bacterial, yeast, or animal cells through the use of recombinant DNA technology, while live vaccines have been produced by introducing relevant genes into the genome of several virus vectors. By using solid–phase peptide synthesis, it has been possible to obtain peptides that mimic the antigenic determinants of viral proteins, that elicit a protective immunity against several viruses. Both the chemical and the recombinant approaches have led to the development of experimental vaccines. It should become clear within a few years which approach will lead to vaccines superior to the ones in use today. The recent development of monoclonal antibody production techniques and enzyme immunoassays permits their application in virology research; diagnosis of viral diseases, and vaccine assessment and standardization. These techniques are useful at different stages in the development of vaccines, mainly in the antigenic characterization of infectious agents with monoclonal antibodies, in assessment schemes in research and clinical assays, and in production.

This new series, entitled Applied Virology Research, is intended to promote the publication of overviews on new virology research data, which will include within their scope such subjects as vaccine production, antiviral chemo– and immunotherapy, diagnostic kits, reagent production, and instrumentation for automation interfaced with computers for rapid and accurate data processing.

We sincerely hope that Applied Virology Research will serve a large audience of virologists, immunologists, geneticists, biochemists, chemists, and molecular biologists, as well as specialists of vaccine production and experts of health services involved in the control and treatment of viral diseases of man, animals and plants. This series will also be of interest to all diagnostic laboratories, specialists, and physicians dealing with infectious diseases.

Edouard Kurstak

Montreal, Canada

Preface to Volume 3

In the first two volumes of this series, emphasis was given to the development of new vaccines and antiviral chemical compounds and to virus variability and the epidemiology and control of viral diseases.

Not long ago, the best that a clinician could hope for in the way of a "rapid" result from a virus diagnostic laboratory was a tentative positive result based on a serological screening test or a immunofluorescent stain preparation. Reports from the laboratory often carried the statement "Further Report to Follow".

Accompanied by a deluge of articles, a revolution in diagnostic methodology has taken place. Based largely on the application of the polymerase chain reaction (PCR) to the detection of viral genomic material, and a multitude of variations of the technique, it is now possible for the diagnostic virology laboratory to issue definitive results on a specimen within a few hours after the specimen has been received. This volume is devoted to the latest diagnostic technology for virus diseases, in particular molecular methodologies.

The editors' appreciation is due to the staff of Plenum Publishing Corporation for their patience and efforts in the production of this series, and in particular, this volume.

E. Kurstak
Montreal, Quebec, Canada
R. G. Marusyk
Edmonton, Alberta, Canada
F. A. Murphy
Davis, California, U. S. A.
M. H. V. Van Regenmortel
Strasbourg, France

Contents

3. The Use of Transcription–based Amplification Systems in the Diagnosis of Human Immunodeficiency Virus HIV-I

Deborah Y. Kwoh and Thomas R. Gingeras

4. Molecular Genetics of Hepatitis C Virus (HCV) and its Use in the Diagnosis of HCV Infections

Edouard Kurstak, Christine Kurstak and A. Hossain

5. Synthetic Peptides in Virology The Use of Synthetic Peptides in Retrovirus Diagnosis

A. Närvänen, M. Korkọlainen, S. Kontio, M–L Huhtala, X. Imai and M. Hayami

6. Fast Multiplex Polymerase Chain Reaction on Boiled Clinical Samples for Rapid Diagnosis of Viral Infections

Christian Vandenvelde and Danièle Van Beers

7. Digoxigenin Labelled Probes and Their Use in the Laboratory Diagnosis of Virus Infections

Pirkko Heino, Veijo Hukkanen and Pertti P. Arstila

8. Biospecific Interaction Analysis in Real Time Using a Biosensor System with Surface Plasmon Resonance Detection

Lars G. Fägerstam

9. Application of Recombinant Antigens for the Diagnosis of Acute Epstein–Barr Virus Infection

Walter Hinderer and Günter Siegl

10. Automation of the Detection of DNA Sequences for the Laboratory Diagnosis of Viral Infections

L. Wray

11. Current Electron Microscopic Methods in Diagnostic Virology

Robert D. O. Devine

Diagnosis Improvements in Virus Infections of Clinical Importance

Christine Kurstak, A. Hossain and Edouard Kurstak

I. INTRODUCTION

Viral diagnosis, until recently, was basically considered to be rather time consuming, fairly expensive and for the most part, inaccessible to the clinician. Currently, however, the role of viruses in infections tend to be increasingly recognized particularly in the immunocompromised patient, the neonate and in sexually transmitted diseases. Furthermore, the recent advances in chemotherapy of viral diseases particularly for serious herpesvirus infections, has made the necessity of a rapid and as well as an accurate viral diagnosis more demanding.

Over the years, many methods have been developed for the laboratory diagnosis of viral infections. Currently, the following are available: virus isolation and identification detection of viral antigens or viral products by immunologic techniques detection of viral genomes by nucleic acid hybridization; direct observation by electron microscopy, and serodiagnosis for detecting antibody response.

II. VIRUS ISOLATION AND IDENTIFICATION

Undoubtedly, the best method for diagnostic would be the isolation and identification of the aetiological agent. However, not all viruses can be grown in cell culture and furthermore, no single cell culture system can be utilized for isolating all or even most of the viruses. Efforts are, therefore, presently being directed towards making available the most sensitive cell culture system for propagating any given virus since virus isolation remains, as yet, the unparalleled method in comparison to other currently available technology. With

C. Kurstak, A. Hossain and E. Kurstak • Department of Microbiology and Immunology, Faculty of Medicine, University of Montreal, Montreal, Quebec, Canada H3C 3J7

Applied Virology Research, Vol. 3, Edited by
E. Kurstak *et al.*, Plenum Press, New York, 1994

the availability of a variety of cell cultures with varying degrees ofsusceptibility to different viruses, it is now possible to separate viral agents based on virus–induced cytopathic effect (CPE). Host cell susceptibility thus, offers an extremely convenient biologic tool for the rapid differentiation of different virus groups. The ability to distinguish poliovirus from coxsackie B viruses in cell culture, prior to performance of serological tests has, for example, led to the establishment of the selective cell culture systems in current use for standard rapid diagnosis of enterovirus infections (Hsiung and Landry, 1986).

Herpes simplex virus (HSV) is the most commonly encountered virus in a clinical virology laboratory. Utilization of an insensitive cell system for its detection may result in a slow positive or even a negative result. Delayed onset of CPE in turn result in a delay in reporting of results by the clinical laboratory. The utilization of cell culture over the past few decades has provided us with a most convenient method for isolation of viruses but unfortunately, very little work towards improving cell culture systems for virus isolation has been forthcoming. It is of importance to continue search for the most sensitive cell culture system for each viral agent. The development of an in vitro system using hepatitis B virus DNA–transfected HepG2 cells (Lampertico *et al.*, 1991), has paved the way towards obtaining effective anti–hepatitis B virus therapeutic agents (Kurstak, 1993).

Improving cell culture conditions to facilitate virus isolation has been dealt with in several studies in the recent past which have noted that centrifugation of clinical specimens onto monolayer may increase the isolation rate and provide an earlier detection of viral isolates (Darouger *et al.*, 1981; Gleaves *et al.*, 1985; Zhao *et al.*, 1987).

Immunologic or hybridization techniques have been applied after amplification of viral infection in cell cultures and often before CPE is evident and thus provides a more rapid and sensitive diagnosis of viral infection. In urine specimens, for example, following centrifugation onto cell culture monolayers, cytomegalovirus was detected at 36 hours post–infection by both immunofluorescence (IF) and immunoperoxidase (IP) techniques, rather than 7 to 14 days needed for detection of CPE (Chou and Merigan, 1982; Gleaves *et al.*, 1984). Tremendous interest has been generated in the commercial kits presently available that incorporate herpes simplex virus (HSV) isolation in cell culture with immunoperoxidase (IP) staining for viral antigen detection (Kurstak, 1986; Kurstak *et al.*, 1986). In addition to a more rapid diagnosis, this combination of methods has the advantage of providing specific identification of HSV, which is of particular importance to laboratories without any existing facilities in virology. Time and materials are also saved since the cultures are terminated at 48 hours instead of the usual 5 to 7 days recommended for conventional HSV cultures (Mayo *et al.*, 1985). An inherent disadvantage is the lack of detection of other viruses which may be present in the specimens (Hughes *et al.*, 1986). A number of studies have compared these commercial kits with conventional isolation in cell culture (Johnson *et al.*, 1985; Hughes *et al.*, 1986). In all, except in two studies (Johnson *et al.*, 1955; Salmon *et al.*, 1986), different cell cultures were used for the two methods. Despite using an inoculum that is three to five fold greater than in routine cell culture isolation, the kits have generally not been as sensitive as conventional culture observed for CPE for 7 to 10 clays. Using the less sensitive Vero cell culture, followed at 48 hours by IP staining, sensitivity of the kit was found to vary from 73% to 79% (Sewell *et al.*, 1984; Hayden *et al.*, 1987). Kits utilizing cell systems other than Vero cells have, however, compared more favourably with conventional virus isolation (Hughes *et al.*, 1986; Salmon *et al.*, 1986). The centrifugation of specimens onto cell monolayers in shell vials has been shown to significantly enhance the rapidity and sensitivity of HSV detection (Johnson *et al.*, 1985; Pruneda and Almanza, 1987; Winter *et al.*, 1987; Zhao *et al.*, 1987). Following preliminary identification of a viral isolate by CPE produced in cell culture, traditional methods as neutralization, hemagglutination–inhibition or complement– fixation tests, utilized for final identification, which are time consuming, have not been less preferred due to the availability recently of fluorescein labeled monoclonal antibodies for influenza A and B, HSV types 1 and 2, adenoviruses and CMV, and polyclonal fluorescein labeled antisera for respiratory syncytial virus (RSV). Utilizing of these procedures for rapid identification could particularly be useful for patient management of those with influenza A during epidemics and for HSV typing in determining the source of the virus infection.

III. DETECTION OF VIRAL ANTIGENS OR VIRAL GENOMES

Cell culture isolation requires technical expertise and is not widely available. Furthermore, many viruses cannot be possibly grown in cell culture. In view of this, interest centered in direct detection of viral antigens by immunologic methods (Kurstak *et al.*, 1986; 1989), and on direct detection of viral genomes by nucleic acid hybridization (Leary *et al.*, 1983; McClure and Perrault. 1986; Gerkin *et al.*, 1991; Choo *et al.*, 1992; Bréchot, 1993).

Among the immunologic methods presently available, immunofluorescence (IF) and immunoperoxidase (IP) techniques have been used for the direct detection of viral antigens in the clinical specimens, including HSV in vesicular lesions (Kurstak *et al.*, 1982; Lafferty *et al.*, 1957), RSV in nasopharyngeal aspirates and swab specimens (Bell *et al.*, 1983; Ahluwala *et al.*, 1987; Jalowayski *et al.*, 1987; Kuman *et al.*, 1987; Ray and Minnich, 1987), and likewise ELISA to detect rotavirus or hepatitis A virus in stool specimens (Coulepis *et al.*, 1985; Cromien *et al.*, 1987).

The spread of diseases such as AIDS, hepatitis B and hepatitis C with their potential for contaminating blood derivatives has created a demand for simple and rapid screening procedures (Hossain *et al.*, 1990). The development of serologic assay to detect antibody to a part (epitope) of hepatitis C virus (anti–HCV) has been a major breakthrough in the long search for a causative agent of non–A, non–B (NANB) hepatitis (Choo *et al.*, 1989, 1992; Kuo *et al.*, 1989). The use of molecular biology techniques for the development of HCV recombinant based antibody assay was a definite deviation from the conventional technology of developing immunoassays by isolation and purifying proteins from the virus itself . The reactive clone produced a recombinant viral antigen (C100–3) that was capable of reacting with antibody in clinical specimens from NANB hepatitis patients and allowed the development of an immunoassay system to detect HCV antibody. In current HCV–ELISA assays, several recombinant HCV antigens (core, 33c, C100, C200, C22) are used, as is PCR to detect HCV RNA in serum of patients, before the appearance of anti–HCV (Cha *et al.*, 1991; Poterucha *et al.*, 1992). Recently, an enhanced detection of HCV infection was reported (Lesniewski *et al.*, 1992) using antigens from the putative polymerase (NS–5) gene product, adding a new tool for diagnosis of hepatitis C.

With the availability of specific therapy for viral infections in particular, rapid diagnosis will gain significant importance in the initiation of appropriate therapy. The DNA probe technology would provide the answer to many of these diagnostic problems. Basic research in the past few years has indeed resulted in the development of technology to produce DNA probes (Leary *et al.*, 1983) consisting of small pieces of DNA which has made it possible for identification and isolation of the genetic information of any organism, and would be particularly useful in the diagnosis of latent or slow virus infections, since the DNA probe detects the presence, rather than the expression of genetic information. DNA probe assays are based on the biochemical principle of DNA double helix with the complementary strands of nucleotides forming stable double stranded molecules. Briefly, the methodology includes the dissolving of the sample to be tested with detergent and enzymes to remove non–DNA components, treating the mixture at a low pH with denatured DNA separating into two strands from the double helix. These strands are bound to a solid matrix and the single–stranded DNA is exposed to an excess of probe DNA which seeks out the complementary sequences on the immobilized DNA and hybridizes to them. This is followed by the washing of the filter which removes the unbound probe from the bound probe that is then detected by the presence of its label (Fucillo *et al.*, 1987). In the past the label was [^{32}P], but presently other labels are replacing radioactive isotopes. A biotin–containing analog now has been incorporated into DNA probes and the biotin detected through the use of avidin–biotin complex. Commercial kits are now available with the essential ingredients for constructing and labeling probes. These assays in combination with immunological reagents have the potential for many rapid viral diagnostic applications. Biotinylated probes, for example, have been used for *in situ* hybridization with intact cells such as paraffin embedded tissues examined for viral genomes, for detection of cytomegalovirus in lung tissue, and the sensitivity of the assay has been found to be close to that of cell culture and IF and monoclonal antibody (Meyerson *et al.*, 1984). Spot hybridization has been utilized to detect cytomegalovirus in urine (Chou and Merigan, 1983) and rotaviruses (Eiden *et al.*, 1987). Nucleic acid hybridization permits viral diagnosis without purification of the pathogen or extensive biochemical or serologic confirmatory tests, and can be highly specific when used in a research laboratory. Concern has been expressed about the specificity of these assays when carried out for the diagnosis of infection. It has been reported that eucaryotic RNA hybridized under stringent conditions with viral RNA from enteroviruses, reoviruses, Sindbis virus and vesicular stomatitis virus (McClure and Perrault, 1985; 1986) and in view of these findings, caution has been advised in the use of nucleic acid probes to detect RNA viruses in clinical specimens. Nevertheless, the use of hybridization techniques with non–isotope labels can offer highly sensitive methods for detection of the presence rather than expression of genetic information, and possess the potential to extend the capability to detect oncogenes and slow viral infections.

It is only a matter of time before technical problems such as limited sensitivity in early or latent infections are solved. A probe direct lesion test for herpes viruses can now identify HSV types 1 and 2 in infected cells in an hour, and it is envisaged that a wide range of DNA probes will be available to the clinical virologist in the near future. A wider application could lead to screening for AIDS and human papilloma virus infection. This new technology will have a significant impact on the screening and diagnosis of sexually transmitted diseases. DNA probes will become a powerful tool in clinical diagnosis and make viral diagnosis a primary rather than confirmatory diagnostic service.

A. *The Polymerase Chain Reaction*

The polymerase chain reaction or PCR is an elegantly developed simple enough technique (Saiki *et al.*, 1985; Mullis *et al.*, 1986). The analytical power of PCR as well as its applicability has been recently demonstrated in the diagnosis and follow–up of patients with hepatitis B infection (Gerkin *et al.*, 1991; Shindo *et al.*, 1991), and for detection of hepatitis C viral RNA (Garson *et al.*, 1990; Weiner *et al.*, 1990). However, nucleic acid probe selection and preparation is technically demanding and their use has not yet eliminated the use of primary culture (Persing, 1991).

In PCR, double–stranded DNA containing a target sequence is mixed in a Tris buffer with all four deoxynucleotides, a thermostable DNA polymerase isolated from *Thermus aquaticus* (Taq polymerase) and at least two oligonucleotides complementary to opposite end of the two strands of the DNA target tissue. The oligonucleotides (primers) anneal to the appropriate DNA strand with a short, double–stranded starting point for Taq polymerase to add nucleotides to the primer and copy of the complementary strand.

PCR is initiated with denaturation of the double–stranded target by a brief incubation (30 sec. to 2 min.) at 92°C to 95°C, rapid cooling to 40°C – 60°C (1–2 min.) to allow the primers to bind to the single–stranded DNA target (annealing) followed by heating to 72°C (30 sec. to 2 min.) for Taq polymerase to elongate the primers. After a single cycle, both original target strands are copied. The power of PCR is achieved through repetitive cycles of denaturation, annealing and primer elongation. As each cycle doubles the amount of target DNA, a million–fold amplification can be accomplished in 30 to 35 consecutive cycles. Following PCR, the sample can be analyzed by a variety of methods, most commonly by electrophoresis on agarose gel and staining with ethidium bromide or another DNA–binding fluorescent dye, then visualizing the amplified material under

UV–light. The identity of post–PCR amplification products is confirmed by traditional hybridization techniques as Southern blotting with radiolabeled oligonucleotides complementary to a region of the amplified DNA or a full–length cDNA probe. The great sensitivity of PCR appears to be a result of exponential amplification of the target sequence with each successive cycle. As demonstrated by Shindo *et al.* 1991) after 30 to 35 cycles, 100 femtogram to 1 picogram of starting target DNA could be easily visualized on ethidium bromide–stained agarose gels. Using a nylon membrane and Southern blotting, the lower limit of detection was reportedly reduced by 100 to 1000 fold down to 100 attogram (Gerkin *et al.*, 1991). The exceptional sensitivity of PCR, however, makes it a necessity to use great care so as to avoid false–positive results. In general, all PCR runs should include positive and negative controls, and the use of a DNA blank subjected to PCR along with the experimental conditions, is essential. PCR by virtue of being a supersensitive technique has enhanced the applicability of modern molecular biological techniques for diagnostic purpose.

B. Direct Observation by Electron Microscopy

Electron microscopy (EM) and immuno–electron microscopy (IEM) have been in use for some time for the rapid identification of viral particles and enabled the discovery of many new viruses in negatively–stained virus–containing clinical specimens. Viruses belonging to the Orthomyxovirus family can be differentiated from the Paramyxoviruses. Rotavirus gastroenteritis, herpes simplex virus, cytomegalovirus, adenovirus, measles virus, hepatitis B virus infections and several other viruses of clinical importance are diagnosed by electron microscopy and results confirmed by immuno–electron microscopy (Kurstak, 1992). The use of protein A for solid phase immuno–electron microscopy is reportedly thirty–fold more sensitive than EM and ten times more sensitive than EIA, as determined with clinical specimens from patients with rotavirus diarrhea (Flewett, 1987).

C. Serodiagnosis for Detecting Antibody Response

IgM antibodies characteristically appear 7 to 10 days after primary infection, reach maximum titers within 2 to 3 weeks, and tend to decline to detectable levels after about 3 months. The presence of IgM specific antibody therefore implies a current or recent infection and offers the potential for a rapid diagnosis with the availability of a single serum specimen.

In pregnancy, maternal IgM antibody does not cross the placenta, hence, the presence of virus specific IgM in the cord blood or serum of a neonate can be reliably assumed to be due to a congenital infection (Hossain and Bakir, 1989).

The value of determining specific IgM antibodies is variable and depends on the particular virus. Transient IgM responses seem to be a characteristic of infections by viruses with antigenic uniformity and eliciting a long–lasting immunity such as rubella, measles and mumps, whereas for others with closely related serotypes such as adeno, entero or parainfluenza viruses with the IgM response being weak or absent, the potential use for diagnosis may be limited. Furthermore, cross–reactions to related viruses of prolonged IgM responses in reactivation of latent viruses as herpes viruses may occur. These variable results and technical pitfalls complicate the interpretation of IgM antibody results. Perhaps, the development of more sensitive methods for the detection of specific IgM may possibly extend the time after infection during which IgM can be detected and thereby extend the diagnostic usefulness of IgM antibody detection.

IV. CONCLUSION

The diagnostic method of choice is largely dependent upon the virus in question and the urgency in obtaining a rapid result. Despite the availability of rapid detection methods for use directly on clinical specimens, cell culture is also performed in order to confirm the specificity and sensitivity of the rapid technique used. Substantial progress has been made in recent years in the rapid and accurate diagnosis of viral infections, but for the laboratory to effectively select the best diagnostic test, the provision of complete clinical data is mandatory as is the proper colection and transportation of the specimens. Above all, increasing and effective communication between the diagnostic laboratory and the clinician concerned along with new, more advanced technologies will play an effective role in making clinical virology truly useful to patient care.

REFERENCES

Ahluwala, G. J., Embree, P., McNicol, B., Low, B. and Hamond, G. W. (1987). *J. Clin. Microbiol.* **25**, 763–767.

Bell, D. M., Walsh, E. E., Kruska, J. F., Schnabel, K. C. and Hall, C. B. (1983). *J. Clin. Microbiol.* **17**, 1099–1103.

Bréchot, C. (1993). *Gut* **34** (Suppl. 2), 39–44.

Cha, T. A.., Kolberg, J., Irvine, B., Stempien, M., Beall, E., Yano, M., Choo, G. L.., Houghton, M., Kuo, G., Han, J.H. and Urdea, M.S. (1991). *J. Clin. Microbiol.* **29**, 2528– 2534.

Choo, Q. L., Kuo, G., Weiner, A. J., Overby, L. R., Bradley, D. and Houghton M. (1989). *Science* **244**, 359–362.

Choo, Q. L., Kuo, G. Weiner, A., Wang, K. S., Overby, L., Bradley, D. and Houghton, M. (1992). *Sem. Liver Disease* **12**, 279–288.

Chou, W. and Merigan, T. C. (1988). *New Engl. J. Med.* **30**, 921–925.

Coulepis, A. G., Veale, M. F., MacGregor, A., Kornitschuh, M. and Gust I.F. (1985). *J. Clin. Microbiol.* **22**, 119–121.

Cromien, J. L., Himmelreich, C. A., Glass, R. I. and Storch, G. A. (1987). *J. Clin. Microbiol.* **25**, 2359–2363.

Darouger, S., Gibson, J. A. and Thaker, U. (1981). *J. Med. Virol.* **8**, 231–234.

Eiden, J., Sato, S. and Yolken, R. (1987). *J. Clin. Microbiol.* **25**, 1509–1512.

Flewett, T. H. (1987). *Post Grad. Doctor* **10**, 41–45.

Fucillo, D. A., Shekarchi, I. C. and Sever, J. L. (1957). In Manual of Clinical Laboratory Immunology (N. R. Rose, H. Friedman and J. L. Pahey, eds.), Amer. Soc. Microbiol., Washington, D.C.

Garson, J. A., Tedder, R. S., Briggs, M., Tuke, P., Glazebook, J. A., Trute, A. and Parker, D. (1990). *Lancet* **335**, 1419–1422.

Gerkin, G., Paterlini, P., Manns, N., Wousset, C., Terve, S., Dienez H.P. and Hess, G. (1991). *Hepatology* **13**, 158–166.

Gleaves, C. A., Smith, T. F., Shuster, E. A. and Pearson, G. R. (1984). *J. Clin. Microbiol.* **19**, 917–919.

Gleaves, C. A., Wilson, D. J., Wold, A. D. and Smith, T. F. (1965). *J. Clin. Microbiol.* **21**, 29–33.

Hayden, F. G., Sorensen, A. S. and Bateman, J. A. (1987). *J. Clin. Microbiol.* **18**, 222–225.

Hossain, A. and Bakir, T. M. F. (1989). *J. Trop. Paediatr.* **35**, 16–20.

Hossain, A., Al–Tuwaijri, A., Kurstak, C. and Kurstak E. (1990). In *Applied Virology Research*, Vol. 2 (E. Kurstak, R. Marusyk, F. Murphy and M. H. V. Van Regenmortel, eds.), pp. 55–59, Plenum Press, New York.

Hughes, J. M., Mann, D. R. and Hamparian, V. V. (1986). *J. Clin. Microbiol.* **24**, 487–469.

Hsiung, G. D. and Landry, M. (1986). *Clin. Microbiol. News* **8**, 173–177.

Jalowayski, A. A., England, B. L., Temm, C. J., Nunemacher, T. J., Bastian, J. F., MacPherson, G. A., Dnakner, N. M., Staube, R. C. and Connor, J. D. (1987). *J. Clin. Microbiol.* **25**, 722–726.

Johnson, P. B., Leavitt, R. N. and Richard, D. F. (1985). *J. Clin. Microbiol.* **21**, 438–441.

Kuman, M. L., Super, D. M., Lambo, R. N., Thomas, F.C. and Prokay, S. L. (1987). *J. Clin. Microbiol.* **25**, 873–878.

Kuo, G., Choo, G. L., Alter, J. H., Getwich, G. L., Redeker, A. H., Purcell, R. H., Nikamura, T., Dienstag, J. L., Alter, M. J., Stevens, C. E., Tegtmera, G. E., Bonino, M., Colombo, M., Lee, N.S., Juo S., Berger K., Shuster, J. R., Overby, L. R., Bradley, D. N. and Houghton M. (1989). *Science* **249**, 362–364.

Kurstak, C., Morisset, R. and Kurstak, E. (1982). *Ann. Virol. Inst. Pasteur* **133E**, 187–190.

Kurstak, E. (1986). *Enzyme Immunodiagnosis*, Academic Press, Orlando.

Kurstak, E., Tijssen, P. and Kurstak, C. (1986). *Bull. W. H. O.* **64**, 465–479.

Kurstak, E., Marusyk, R., Salmi, A., Babiuk ,L., Kurstak, C. and Van Regenmortel, M. (1989). In *Subcellular Biochemistry, Vol. 15, Virally Infected Cells,* (J. R. Harris, ed.), Plenum Press, New York.

Kurstak, E. (1992). *Control of Virus Diseases*. Second Edition. Marcel Dekker Inc., New York.

Kurstak, E. (1993). *Viral Hepatitis: Current Status and Issues.* Springer–Verlag, Wien and New York.

Lafferty, W. E., Krofft, S., Remington, M., Giddings, R., Winter, C., Cent A., Corey L. (1987). *J. Clin. Microbiol.* **25**, 323–327.

Lampertico, P., Walter, J. S. and Gerber, M. A. (1991). *Hepatology* **13**, 422–426.
Leary, J. J., Brigati, D. J. and Wood, D. C. (1983). *Proc. Natl. Acad. Sci. USA.* **80**, 4045–4049.
Lesniewski, R. R., Desai, S. M., Johnson, R. G., Nelson, L. R., Schlauder, G. G., Sant, C. L. and Mushahwar, I. K. (1992). *Proc. European HCV Council Group Meeting*, Venice.
Mayo, D. R., Breman, T., Egverston, S. H., Moor, D. F. (1985). *J. Clin. Microbiol.* **21**, 768–773.
McClure, M. A. and Perault, J. (1986). *J. Virol.* **57**, 917–921.
Mullis, K., Faloona, F., Scharf, S., Saiki, R., Horn, G. and Erlich, H. (1986). *Cold Spring Harbor Symp. Quant. Biol.* **51**, 263–273.
Myerson, D. R., Hackman, R. C. and Meyers, J. D. (1984). *J. Infect. Dis.* **150**, 272–276.
Persing, D. H. (1991). *J. Clin. Microbiol.* **29**, 1251–1285.
Poterucha, I. J., Rakela, J., Lumeng, L., Lee, C. H., Taswell, H. F. and Wiesner, R.H. (1992). *Hepatology* **15**, 42–45.
Pruneda, R. C. and Almanza, I. (1987). *J. Clin. Microbiol.* **25**, 423–428.
Ray, C. G. and Minnich, L. L. (1987). *J. Clin. Microbiol.* **21**, 286–290.
Saiki, R., Scharf, S., Faloona, P., Mullis, K., Horn, G., Erlich, H. A. and Arnheim, N. (1985). *Science* **230**, 1350–1354.
Salmon, V. C., Turner, R. B. and Speranza, M. J. (1986). *J. Clin. Microbiol.* **23**, 683–657.
Sewell, D. L., Horn, S. A., Dilbeck, P. M. (1984). *J. Clin. Microbiol.* **19**, 705–709.
Shindo, M., Okuno, T., Arai, K., Matsumoto, M., Takeda, M., Kashima, K. and Shimada, K. (1991). *Hepatology* **13**, 167–171.
Weiner, A. J., Juo, G., Gradley, D. M., Bonino, F., Saracco, G., Lee, I. and Rosenblatt, J. (1990). *Lancet* **335**: 1–3.
Winter, G. F., Inglis, J. M., Cubie, H. A. (1987). *J. Virol. Methods* **15**, 329–335.
Zhao, L., Landry, M. L., Balkovic, E. 5., Hsiung, G. D. (1957). *J. Clin. Microbiol.* **25**, 1401–1405.

2

Diagnosis of HIV Infection and the Polymerase Chain Reaction (PCR)

Gerald Schochetman and John J. Sninsky

I. INTRODUCTION

Since the description of the first cases of acquired immunodeficiency syndrome (AIDS) in 1981, the global impact of this new epidemic has been dramatic. By July, 1993, more than 150 countries had reported over 718,894 cases to the World Health Organization (WHO, 1993), and more than 8 million persons worldwide were estimated to be infected with the human immunodeficiency virus type 1 (HIV–1) , the etiologic agent of AIDS. There are estimated to be at least 1 million HIV infected persons in the United States alone. Most of these infected persons are expected to develop AIDS. HIV–1 was first isolated in 1983–1984 from patients with ARC and AIDS (Barre–Sinoussi *et al.*, 1983; Gallo *et al.*, 1984; Popovic *et al.*, 1984) and is found worldwide, primarily in Central Africa, Europe, North and South America and now in Asia. A second AIDS virus HIV–2 (Clavel, 1987; Clavel *et al.*, 1986; Horsburgh and Holmberg, 1988), closely related to the simian immunodeficiency virus (SIV) and more distantly related to HIV–1, was discovered in 1986 and shown to be endemic in parts of West Africa and to have limited spread in Western Europe, the United States, Canada, and Brazil.

II. TESTING FOR HIV

Virus isolation in PHA–stimulated lymphocytes and successful adaptation to growth in continuous T–cell lines made possible the development of tests to detect antibodies to HIV–1 in infected persons. This led to the ability to identify both symptomatic and asymptomatic carriers of the virus. The first commercial tests in the U.S., enzyme immunoassays (EIA), were licensed by the Food and

---

Gerald Schochetman • Division of HIV/AIDS, Centers for Disease Control United States Public Health Service, Department of Health and Human Services, Atlanta, Georgia. *John J. Sninsky* • Department of Infectious Diseases, PCR Division, Cetus Corporation, Emeryville, California.

Drug Administration (FDA) in 1985. These tests were immediately used fordonated blood and plasma screening, and subsequently for general diagnostic use. Because any screening test can result in a false positive reaction, it is necessary to validate positive results by supplemental testing. This has mainly been accomplished using Western blots (WB) or to a lesser extent, indirect immunofluorescence assays (IFA) .

Tests for HIV have been used for three main purposes, namely screening, diagnosis, and surveillance. Efforts to screen for HIV infection have depended primarily on the EIA because it is highly sensitive and specific and because the EIA format is well-suited for mass testing. Although initially developed to screen blood donors, the tests have also been used to screen many other groups including hospitalized patients, military recruits and personnel, prisoners, applicants for marriage licenses, applicants for health and life insurance, child-bearing women and their newborns, and immigrants.

Medical diagnostic testing includes detection of HIV infection, and staging and prognosis of HIV disease. Antibody screening by EIA followed by confirmation using WB or IFA remains the most widely used method of determining HIV infection. However, diagnosis of infection cannot always be based on antibody tests. Antibodies are not present very early in the infection and may be lost in advanced AIDS.

Patients with congenital infections or immune deficiencies may not seroconvert (develop detectable virus-specific antibodies). Furthermore, antibodies can be passively acquired as occurs with maternal antibodies in neonates born to HIV seropositive mothers.

Although only a small percentage of these neonates are truly infected virtually all are seropositive due to the acquisition of maternal antibodies. In these situations, use has been made of other approaches to determine HIV infection including direct detection of the virus or its components by a) culture, b) analyzing plasma or serum for circulating viral antigen, and c) the use of the polymerase chain reaction (PCR) technique for the presence of viral genetic information. For perinatal diagnosis, use has recently been made of assays for cell surface HIV antibody expression (ELISPOT test) or *in vitro* HIV antibody production (IVAP) by the patient's lymphocytes.

III. BIOLOGY OF HIV

HIV-1 and HIV-2 are members of the retrovirus family of viruses whose infectious viral particles contain single-stranded RNA (ssRNA) as their genetic information. It was observed early in the AIDS epidemic that the T4 helper lymphocyte showed a dramatic decline in infected patients. It was subsequently shown that the T4 cells were the principal target of HIV-1 and that their loss was due to their preferential infection and destruction by the virus. The preferential infection of T4 cells by HIV-1 was due to the presence of the CD4 protein on the surface of these cells. It was later discovered that monocytes and macrophages, which also express the CD4 protein, could likewise be infected and become functionally compromised. During the HIV life cycle, the viral ssRNA genome undergoes conversion to a double–stranded DNA (dsDNA) form using the viral enzyme reverse transcriptase (RT). This dsDNA, termed the provirus, may become integrated into the host cell's chromosomes and remain associated with the cellular chromosomal DNA for the life of the infected cell.

Following integration, the provirus can follow either of two pathways. The provirus could either undergo active transcription of the genes coding for the structural proteins of the virus resulting in the assembly and release of infectious virions, or it could undergo selective transcription of only the complex array of viral regulatory genes, remain transcriptionally restrained, and not release viral particles. This latter condition is frequently referred to as the "latent state. " Since the HIV proviral DNA is present regardless of the transcriptional state of the cell, initial efforts for direct detection of the virus used proviral DNA as a template.

Because the number of HIV–1 infected peripheral blood mononuclear cells (PBMCs) in an infected person are low (Harper *et al*., 1986), established molecular biology techniques (Shaw *et al.*, 1984) proved insensitive to routinely detect and characterize HIV proviral DNA directly from a patient's PBMCs. Overcoming this problem required amplifying HIV–specific DNA to detectable levels using the polymerase chain reaction (PCR). Before the advent of PCR, direct detection of HIV–1 was achieved by virus coculturing. PCR has been used to selectively amplify HIV sequences orders of magnitude *in vitro,* and has eliminated the need for lengthy coculturing of the virus for its direct detection.

PCR has proven to be extremely important in the study of HIV infection and AIDS, and has demonstrated both clinical and research utility for: (a) the direct detection and quantitation of HIV DNA and RNA from cells of infected persons; (b) detecting infected persons during the window period, i.e., prior to the generation of HIV specific antibodies); (c) resolving the infection status of individuals with an indeterminate Western blot; (d) screening of neonates for HIV infection; (e) distinguishing HIV–1 from HIV–2 infections; and (f) defining the patterns of transmission and evolution of the virus throughout the population.

IV. PCR TECHNOLOGY

PCR was devised as a method for the in vitro amplification of selected DNA sequences (Mullis *et al*., 1987; Saiki *et al.*, 1985; Saiki *et al*., 1988; Schochetman *et al.*, 1988). For detection of HIV proviral DNA, PCR sample preparation requires the separation of mononuclear cells from the polymorphonuclear cells of a blood specimen using a Ficoll–hypaque gradient. Following the preparation of the peripheral blood mononuclear cells (PBMC) , the cells are incubated with non–ionic detergents followed by treatment with proteinase K to liberate the cellular DNA. The proteinase K is then inactivated by heating the sample to 90°C for 10 minutes and the resulting DNA preparation is ready for PCR. One μg of DNA (equivalent to about 150,000 mononuclear cells; see Table 1) is routinely used per PCR reaction.

Table 1. Numbers to Consider for HIV Infection

Whole Blood

- 5.6 liters of blood in a 70 kg person
- 500,000 white blood cells per 65 μl whole blood

White Blood Cells

- 350,000 polymorphonuclear cells (70%)
- 150,000 mononuclear cells (30%)

Mononuclear Cells

- 15,000 monocytes (10%)
- 15,000 large granular lymphocytes (LGL) (10%)
- 15,000 B–cells (10%)
- 105,000 T–cells (70%)

T–Cells

- 37,000 T8 (suppressor) cells (35%)
- 68,000 T4 (helper) cells (65%)

PCR is a repetitive process consisting of three distinct steps (see Fig. 1). These steps include, (a) denaturation of the dsDNA, (a) annealing of specific primers and (c) extension of the annealed primers. Due to the complementary and anti–parallel nature of DNA, ssDNA can also serve as a template for amplification. If a specific RNA sequence is to be amplified, a DNA copy (cDNA) of the RNA sequence is produced using the enzyme reverse transcriptase (RT) prior to PCR amplification of the resulting cDNA (Hart *et al.*, 1988; Byrne *et al.*, 1988). Following PCR amplification, a number of techniques can be used to detect the amplified DNA sequences. The resulting amplified DNA is of a defined size and can sometimes be visualized after gel electrophoresis. However, this method cannot provide definitive identification of the product. Confirmation of amplification of HIV DNA requires hybridization of an aliquot of the amplified DNA to a synthetic DNA probe which is complementary to a portion of the amplified DNA sequence. The probe can be labeled by a variety of means, isotopic (radioactive) or nonisotopic (colorimetric or chemiluminescent; see Figures 2–4) .

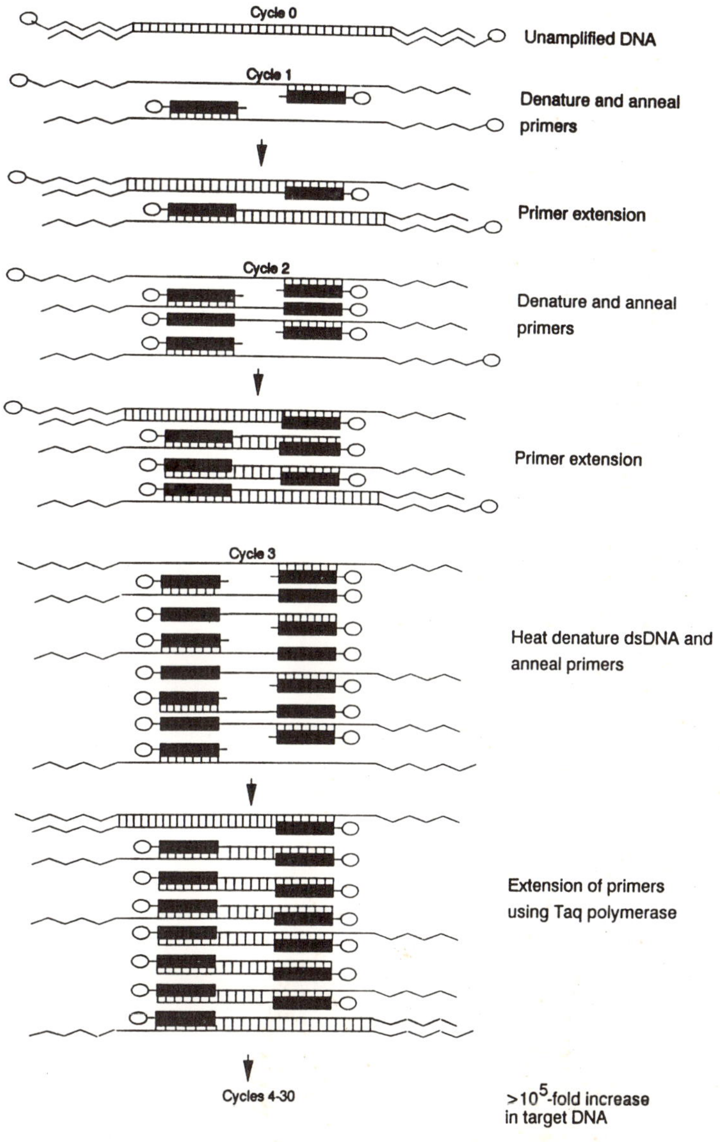

Figure 1.
Polymerase chain reaction (PCR). PCR is a repetitive process including denaturation of double–stranded DNA (dsDNA), annealing of primers, and extension of bound primers. One PCR cycle usually takes only a few minutes, and the cycle is repeated many times, typically 25–35 times. The dsDNA is first heated to 95°C to separate the strands of the DNA duplex. During the subsequent annealing phase, oligonucleotide primers hybridize to the dissociated HIV DNA. Each primer is complimentary to one of the original DNA strands, to either the 5' or 3' side of the sequence of interest. Following annealing, a thermostable DNA polymerase from *T. aquaticus* (Taq) is used to catalyze the synthesis of new strands of DNA that are complementary to the intervening sequences primed by the opposing oligonucleotide primers.

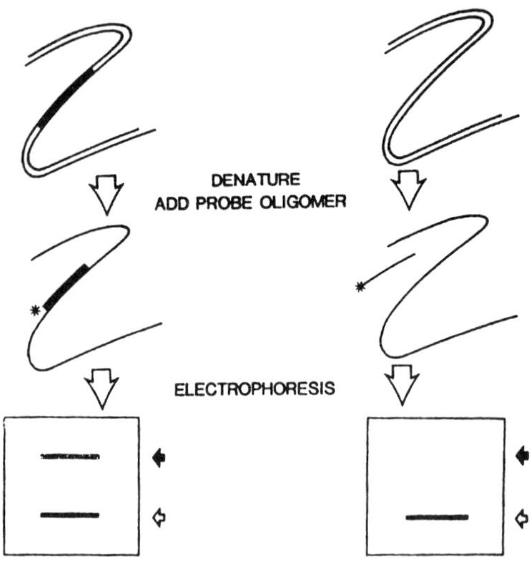

Figure 2.
Amplified DNA detection involves hybridization of a portion of the amplified DNA product to a radioactive [^{32}P]–labeled synthetic probe complementary to a portion of the amplified sequences followed by gel analysis and autoradiography.

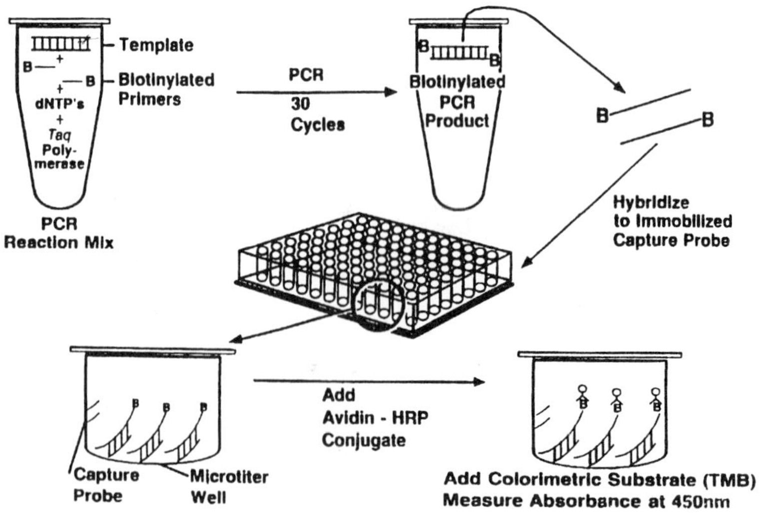

Figure 3.
Detection of the amplified DNA involves the use of biotinylated primers to generate a tagged amplified DNA product which is then hybridized to an immobilized probe complementary to a portion of the amplified sequences followed by incubation with avidin conjugated to horse radish peroxidase This standard EIA format in a microplate yields a colorimetric readout for positive samples (C. Silver, M. Sulzinski, E. Dragon, and M Longiaru, *personal communication*).

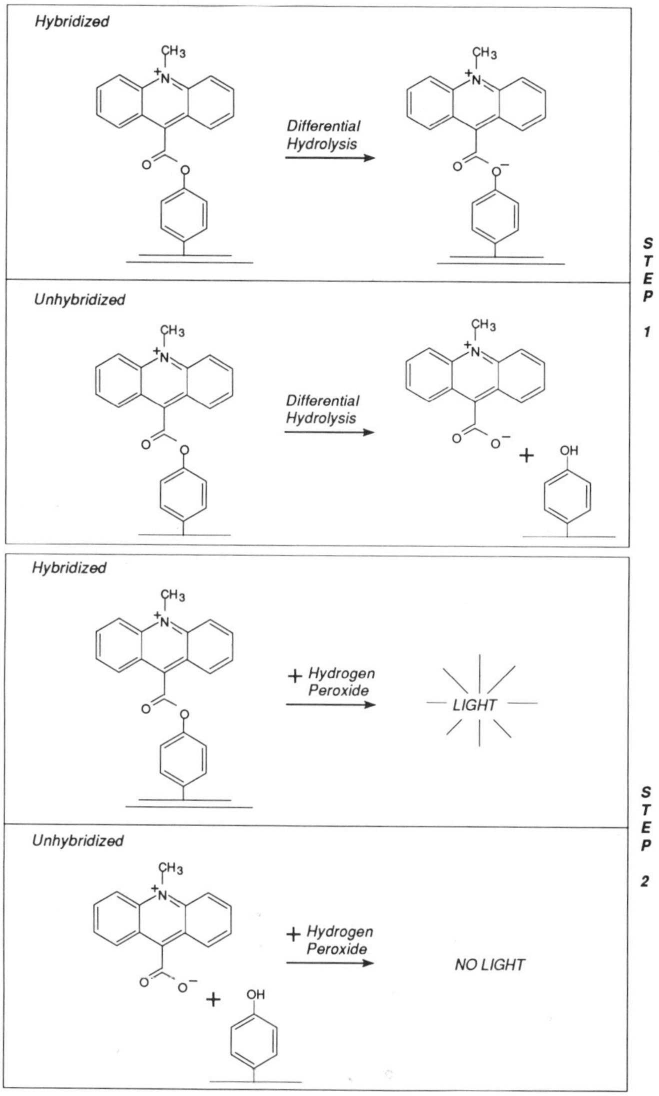

Figure 4.
Detection of the amplified DNA involves hybridization of a portion of the amplified DNA product to an acridinium-labeled synthetic probe complementary to a portion of the amplified sequences followed by differential alkaline hydrolysis. The acridinium attached to the hybridized probe is relatively resistant to alkaline and can chemiluminescence after oxidation by the addition of hydrogen peroxide. In contrast, the acridinium attached to the unhybridized probe is highly sensitive to alkaline degradation and loses its ability to chemiluminescence almost immediately.

V. HIV PROVIRAL DNA DETECTION

PCR has been successfully used for direct detection of HIV–1 proviral DNA sequences in PBMCS of seropositive persons (Ou *et al*., 1988; Kwok *et al*, 1987; Kwok *et al*., 1989; Kellogg and Kwok, 1989) and to detect HIV–1 DNA in cells from seropositive subjects who were negative by virus coculture (Ou *et al*., 1988). Recent studies have demonstrated that PCR has simplified the ability to directly clone and sequence HIV–1 DNA (Ou and Schochetman, 1989) and HIV–1 cellular RNA (Hart *et al*., 1988) allowing for a more accurate assessment of the repertoire of HIV sequences present within a patient (Meyerhans, *et al*., 1989). This results from the fact that PCR does not require virus isolation which results in the selection of only a subset of the HIV strains that are present in a patient (Meyerhans *et al*., 1989; Goodenow *et al*., 1989) .

PCR has been used to demonstrate that essentially all HIV antibody–positive persons are infected with the virus further confirming that HIV is the causative agent of AIDS (Jackson *et al*., 1990). This large scale study involved testing of PBMCS obtained from 409 persons who were HIV antibody positive. Among these seropositive persons, 56 had AIDS, 88 had ARC and 265 were asymptomatic. Blood samples from an additional 131 persons who were HIV antibody negative were also examined. It was reported that the 56 AIDS patients, 87 of the 88 ARC patients (99%) , and 259 of the 265 asymptomatic but HIV antibody positive–persons (98%) tested positive for virus by culture, by PCR, or by both analyses; in contrast, none of the 131 HIV–1 antibody negative persons was positive for virus by culture or by PCR.

VI. HIV PROVIRAL DNA QUANTITATION

PCR can be used to quantitate the number of HIV infected cells (Kellogg *et al*., 1990) or the amount of cell–free virus in the plasma (viremia) of a patient by determining the amount of original target DNA. A quantitative assay for proviral HIV–1 would be important for the evaluation of drug and vaccine efficacy, or for monitoring disease progression. Quantitation of viral burden in infected persons has been performed with varying success, by limiting dilution culture methods in which culture supernatants were tested either for the presence of HIV–1 gag protein (p24) or the presence of viral RT activity or both (Ho *et al*., 1989; Coombs *et al*., 1989). Unfortunately, culturing virus is expensive, time consuming, requires handling of large volumes of infectious material, and has not always been reliable. By amplifying a dilution series of known amounts of HIV DNA (e.g. a plasmid containing a full length copy of HIV–1 DNA, or DNA from a cell line containing one integrated copy of HIV–1 per cell) , it has been possible to quantify the virus burden in a person by determining the number of HIV proviral copies per known number of cells in that person (Kellogg *et al*., 1990; Ou *et al*., 1990; Schnittman *et al*., 1990, Ratner, 1989; Lion *et al*., 1989; Schnittman *et al*., 1990; Simmonds *et al*., 1990; Spear *et al*., 1990). To ensure the biosafe use of a full–

length proviral DNA standard, a replication–deficient HIV–1 proviral DNA has been developed (Hart *et al*., 1990). A cell line, ACH2, containing one integrated copy of HIV–1 and which produces a non–infectious HIV has also been developed (Clouse *et al*., 1989). Quantitation of specific RNAs by PCR has also been reported (Wang *et al*., 1989; Becker–Andre and Hahlbrock, 1989; Gilliland *et al*., 1990). Recently, a rapid and quantitative procedure was reported for detecting HIV–specific amplified DNA using a nonradioactive acridinium–labeled DNA probe which chemiluminesces to give an optical readout (Ou *et al*., 1990). The total time for PCR amplification followed by DNA probing using this new technique was about 4 hours. Thus, detection and quantitation of HIV DNA could be achieved in 1 to 1.5 days from the time of receipt of the blood sample.

The relation between infected cell burden and immunologic status in persons with asymptomatic and symptomatic HIV infection has been reported by a number of investigators using PCR (Schnittman *et al*., 1990, Schnittman *et al*., 1990; Simmonds, *et al*., 1990; Spear *et al*., 1990). In those studies, DNA from PBMCS from different patients were serially diluted, amplified, and the presence of HIV–specific DNA sequences were determined using a radiolabeled DNA probe. The intensity of the signal from each amplification was compared with PCR performed on serial dilutions of the plasmid containing the HIV–1 genome or the ACH2 cell line that contains one integrated copy of HIV–1 per cell.

These studies demonstrated that there was a significant increase in viral burden per constant CD4+ cells in patients as they progressed to clinical disease along with a concomitant quantitative depletion of CD4+ cells. This contrasted with the stable viral burden and the maintenance of a relatively constant level of CD4+ cells in patients who were clinically stable (Schnittman *et al*., 1990). The number of HIV infected cells can be calculated from the number of proviral copies based on the reported estimate that there is approximately 1 proviral copy per infected CD4 cell (Simmonds *et al*., 1990). These results can be compared to the number of cells producing virus in asymptomatic versus symptomatic persons determined by limiting dilution cultures (Ho *et al*., 1989; Coombs *et al*., 1989). Based on this type of analysis it can be estimated that approximately 10% of the infected cells in asymptomatic persons are actively expressing virus, compared to virtually I00% of the infected cells in symptomatic persons. These results indicate that as patients progress from an asymptomatic to a symptomatic state the number of infected cells and the proportion of infected cells actively expressing HIV increases substantially. This is consistent with a direct and probable causal relation between increase in viral burden, and immunosuppression and disease, presumably do to the increase in HIV expression leading to eventual destruction of CD4+ cells.

Cells undergoing active transcription can be differentiated from those in a latent state (Hart *et al*., 1988, Byrne *et al*., 1988, Murakawa *et al*., 1988) because HIV–specific RNA sequences can be detected in cells of infected persons by amplifying cDNA copies of reverse–transcribed cellular RNA. The assay was reported capable of detecting HIV RNA in 1 infected cell among 10^6 uninfected cells (Hart *et al*., 1988). Comparison of the presence of detectable HIV serum antigen, with HIV–specific RNA expression in the same patients (Hart *et al*., 1988) demonstrated that RNA PCR was more sensitive than serum antigen detection (i.e., all patients that were antigen positive were HIV RNA positive, but all patients that were HIV RNA positive were not antigen positive).

VII. HIV INFECTION AND SEROCONVERSION

A number of studies have indicated that in some persons HIV can be detected by virus isolation, antigen detection or PCR prior to seroconversion (Goudsmit *et al.*, 1986; Kessler *et al.*, 1987; Ranki *et al.*, 1987; Simmonds *et al.*, 1987; Ward *et al.*, 1989; Loch and Mach, 1988; Wolinsky *et al.*, 1989; Hewlett *et al.*, 1988; Imagawa *et al.*, 1989). Some have reported cases of positive antigen reactions for 14 Months or More without detectable antibody (Ranki *et al.*, 1987; Wolinsky *et al.*, 1989).

There have been reports of either positive cultures or positive PCR reactions for extended periods prior to seroconversion (Wolinsky *et al.*, 1989, Imagawa *et al.*, 1989). An initial attempt to duplicate one of these studies (Imagawa *et al.*, 1989) has been unable to verify the original results (unpublished data) and will require further investigation. Individuals who have been followed prospectively after exposure to HIV–1 usually seroconverted within 6 months (Horsburgh *et al.*, 1989). To define the length of time from infection to the ability to detect HIV–specific antibodies, a study was conducted of 26 homosexual men and 11 men with hemophilia pre– and post–seroconversion (Horsburgh *et al.*, 1989). PBMCS from these men were analyzed for HIV–1 DNA by PCR using primers from the env and gag regions of the viral genome. Using a Markov statistical model, it was estimated that the median time from infection with HIV–1 to seroconversion was within 3 months and that 95% of all persons who became infected would seroconvert within 6 months. These results indicated that lengthy periods of latent infection without detectable antibody probably are rare.

VIII. PCR FOR PERSONS WITH INDETERMINATE WESTERN BLOTS

PCR has been used to determine whether apparently healthy persons who had repeatedly reactive enzyme immunoassays (EIAS) and an indeterminate Western blot were infected with HIV–1 (Jackson *et al.*, 1990). Sera from 99 volunteer blood donors, from a HIV–1 low prevalence area of the United States, with the above serologic outcome were coded and tested for the presence of HIV by culture and PCR. Ninety–eight of the 99 blood donors had no identified risk factors for HIV–1 infection whereas one donor had used intravenous drugs. After a median of 14 months from the time of the initial serologic testing, 65 donors (66%) were still repeatedly reactive for HIV–1 on at least one immunoassay. In 91 donors (92%), the Western blot results remained indeterminate. None of the 99 donors had evidence of HIV infection as determined by both culture or by PCR.

These results demonstrate that infection in persons at low risk for HIV–1 infection, such as volunteer blood donors, and who have persistent indeterminate HIV–1 Western blots are rare.

IX. HIV–1/2 TYPING

HIV–1 and HIV–2 exhibit serologic cross–reactivity between their gag (core) proteins in contrast to their env (envelope) proteins where cross–reactive antibodies are thought to be much less common. Recently, there have been reports of persons who were seroreactive against the gag, pol and env proteins of both HIV–1 and HIV–2 (George *et al.*, 1989; Rey *et al.*, 1986; Rey *et al.*, 1987; Foucault *et al.*, 1987; Rayfield *et al.*, 1988) .

By serologic means it has been difficult to determine whether this dual seroreactivity was due to: a) infection with either one of the HIVs eliciting a broad immune response to epitopes common to both viruses; b) an infection with a recombinant or third virus containing determinants of both HIV–1 and HIV–2; or c) a true mixed infection with both viruses in the same person. In regions where HIV–1 but not HIV–2 is highly endemic, such as the United States, it would be unlikely to find a person infected with both viruses. However, in certain areas of West Africa, such as the Ivory Coast where HIV–1 and HIV–2 are both prevalent, the probability of finding a person infected with both viruses would be enhanced.

PCR has been used in conjunction with serology to determine if a patient was infected with HIV–1, HIV–2 or both viruses (Rayfield *et al.*, 1988). The technique has been successfully employed to confirm the first case of HIV–2 infection in a person living in the United States (SS) and has been also use to confirm the first case of a mixed HIV–1 and HIV–2 infection in the same person (Rayfield *et al.*, 1988). This latter person was seroreactive by whole virus EIAS, type–specific peptide EIAS, and Western blots for both viruses. In addition PBMCS from this person contained proviral sequences of both HIV–1 and HIV–2 as determined by PCR.

X. HIV INFECTION IN INFANTS

HIV transmission via blood and blood product transfusions has virtually been eliminated in mothers developed countries because of donor deferral and screening of the blood supply. Thus, perinatal or vertical transmission currently accounts for nearly all incident HIV infection in prepubertal children. The rate of transmission from mothers to infants varies among different studies. Although results from the larger, multicentre prospective studies indicate that 20% to 40% of infants born to women with HIV infection will acquire the virus from their mothers, recently reported studies have shown lower rates, below 20% (Scott *et al.*, 1985; European Collaborative Study, 1991; Blanche *et al.*, 1989; Italian Multicentre Study, 1988).

The laboratory and clinical diagnosis of HIV–1 infection in children presents some unique problems. Because of passive transfer of maternal antibody in utero, HIV–1 infection in infants born to HIV infected mothers cannot be confirmed with the serologic assays commonly used for diagnosis in adults. Certain clinical features such as disease manifestations, and incubation and latency period to development of AIDS, are different in children compared with adults. Understanding these differences is important for diagnosing HIV infection in children.

PCR appears to be quite promising for early diagnosis and possibly prognosis of HIV infection in infants born to HIV–infected women (Rogers *et al.*, 1989). PCR has advantages that highlight its usefulness as an early and rapid diagnostic test for newborns. These include: a) its ability to detect HIV proviral DNA and therefore avoid the problem of maternal antibody; b) it can be performed on 2–3 ml of blood, a volume that can practically be obtained from a newborn; c) it is rapid and can be completed in one days time; and d) it is less biohazardous than HIV culture.

PCR has been used successfully to diagnosis HIV–1 infection by detecting HIV–1 DNA during the neonatal period (0 to 28 days of age) in those infants born to HIV seropositive mothers who develop a severe and rapid course of disease (Rogers *et al.*, 1989; Rogers *et al.*, 1990). One study of serial samples collected from infants followed during the first 2 years of life found that approximately 50%–60% of HIV–infected infants were PCR positive during the neonatal period and virtually 100% were positive at some time during the postneonatal period (Rogers *et al.*, 1989; Rogers *et al.*, 1990). Most infants were found to be PCR positive by 6 months of age. Infants who developed AIDS during the first 18 months of life were more likely to be PCR positive in the neonatal period compared with infected infants who had milder course of the disease. Other studies have found similar results when testing infants during the postneonatal period, but few have tested significant numbers of infants in the neonatal period (Krivine *et al.*, 1990; Chadwick *et al.*, 1989).

It is currently unclear why infected infants may be PCR negative in the neonatal period. It may be that infected infants are PCR negative in the neonatal period if transmission occurred late in gestation or during labor and delivery. Alternatively, transmission could take place in utero but the level of proviral DNA at birth is below the level detectable by the procedure using the size sample tested to date.

PCR is also useful in documenting lack of HIV infection in infants who lose their maternal antibodies. In a study of 119 infants who lost maternal antibody, none had conclusive evidence of HIV infection by PCR (Abrams *et al.*, 1990). Six of the 119 infants tested PCR positive on one occasion, but none could be confirmed by repeat PCR testing, or other laboratory procedures or clinical examination. It was suspected that the single positive PCR tests were false positive results due to either shipping or handling problems, or to laboratory errors. In fact, HLA typing of discordant samples from the 6 children indicated mislabeling in at least 2 children since discordant samples had different HLA types. Alternatively, these children may have been truly infected but with levels of HIV below the limits of sensitivity of the test in most samples. The continued development of highly sensitive and specific tests for detecting HIV infection will be necessary to allow for better understanding of when the virus is transmitted from mother to infant and for detecting it in infants early in the course of their illness.

Ideally, a laboratory test should diagnose infected infants at the time of birth to allow for the earliest opportunity to intervene with antiviral (Pizzo *et al.*, 1988) or other prophylactic therapy.

XI. MONITOR DRUG-RESISTANT HIV BY PCR

Zidovudine™ or azidothymidine (AZT) has demonstrated the ability to extend life expectancy and to lower the frequency and severity of opportunistic infections. The first isolates reported with reduced sensitivity in vitro to Zidovudine™ (Larder *et al*., 1989) , were cultured from patients receiving the drug, and later found to contain specific mutations within the coding sequence for the HIV RT (Larder and Kemp, 1989). It has been demonstrated that the HIV isolates most resistant to Zidovudine™ have four amino acid substitutions at positions 67, 70, 215, and 219; whereas isolates that have a subset of these four mutations are less resistant to the drug. Recently, a "nested" or "double" PCR procedure was developed (Boucher *et al*., 1990) to detect the common mutations found in residue 215 (e.g. conversion of the threonine codon to one for either tyrosine or phenylalanine requires a two base change). By modifying the PCR conditions (Kwok *et al.*, 1990) , PCR can be used to selectively amplify sequences varying by a single nucleotide making the other codon changes amenable to a similar analysis. The role these mutations play in the declining efficacy of Zidovudine™ after long periods of treatment remains unclear; however, rapid and sensitive diagnostic procedures for their detection will aid in resolving their contribution.

XII. ANALYTICAL SENSITIVITY AND SPECIFICITY

PCR exhibits the best analytical sensitivity of any test that has been employed in the laboratory diagnosis of infection with a pathogenic agent. Using PCR, single molecule detection has been reported by numerous laboratories. This superb analytical sensitivity doesn't necessarily translate into diagnostic sensitivity and subsequently clinical utility. PCR doesn't have an intrinsic analytic or diagnostic sensitivity and specificity. The diagnostic sensitivity and specificity is linked to the laboratories performing the procedure. Thus, the confidence in the reported results is directly proportional to the experience and to the critical interpretive criteria used by the laboratories performing the assay.

Multiple parameters have been shown to dramatically effect the overall analytical sensitivity of PCR and correspondingly its diagnostic sensitivity. Factors with demonstrated ability to effect the overall amplification efficiency beyond the obvious contribution of the selected primers and probes for amplification and detection, respectively include the concentration of the various reagents and the thermocycling profiles used for amplification .

Therefore, as with other diagnostic assays, the use of well characterized controls to monitor inter– and intra–assay variability is vital. Unfortunately, the use of PCR for the detection of HIV has resulted in several controversial reports that contradict the experience of the remainder of the diagnostic community carrying out this procedure. PCR data in the absence of patient follow–up and supporting results from more established procedures such as the FDA approved EIAS and Western blots, or virus coculture by an experienced laboratory should be viewed with caution.

As with other diagnostic assays, the use of PCR for the detection of HIV has yielded both false–positive and negative results. False positives have been demonstrated to occur because of (a) cross contamination of a negative sample from a positive sample, (b) contamination of clinical samples or the reagents for amplification with recombinant plasmids or phage harboring the entire or portions of the HIV proviral genome, and (c) "carryover" of PCR products from previous positive reactions. The latter is usually the reason for false–positives because of the number of copies generated by PCR , e.g., $10^6 - 10^{12}$. Higuchi and Kwok (1989) have recommended specific precautions to follow in order to minimize this type of contamination (see Carry Over below). Recently, two laboratories (Longo *et al*., 1990) have described the use of dUTP instead of dTTP and the other three conventional deoxynucleoside triphosphates in PCR and pretreatment of all reactions with uracil DNA glycosylase to eliminate or "sterilize" errant PCR products as templates for the amplification (Figure 5) .

Just as PCR harnesses the replication capacity of cells, this procedure exploits the restriction/modification and excision/repair systems of cells. Since PCR products containing du hybridize as efficiently as dT containing PCR products, and can be cloned and sequenced, this procedural modification promises to increase the reliability of positive results from a larger number of laboratories.

False–negatives have been attributed to compromised analytical sensitivity because of insufficient specificity either because of less than optimal amplification conditions or the selection of primers and probes that do not readily recognize different sequence variants (Kwok *et al*., 1990) .

As with all diagnostic assays, replication is an important element to be confident that the results are reproducible. Samples that have disparate results in duplicate, not unlike the discordant EIA assays, can be caused by signals at the cutoff for positivity or by sample mixup. A sample must contain five copies of HIV template to have a 99% likelihood of being reproducibly positive. If there are fewer than an average of five copies in a sample, the reactions may appear irreproducible due to sampling bias.

Recently, a multicentre, blinded proficiency trial using 105 HIV–1 seronegative, culture negative samples from low risk blood donors and 99 HIV–1

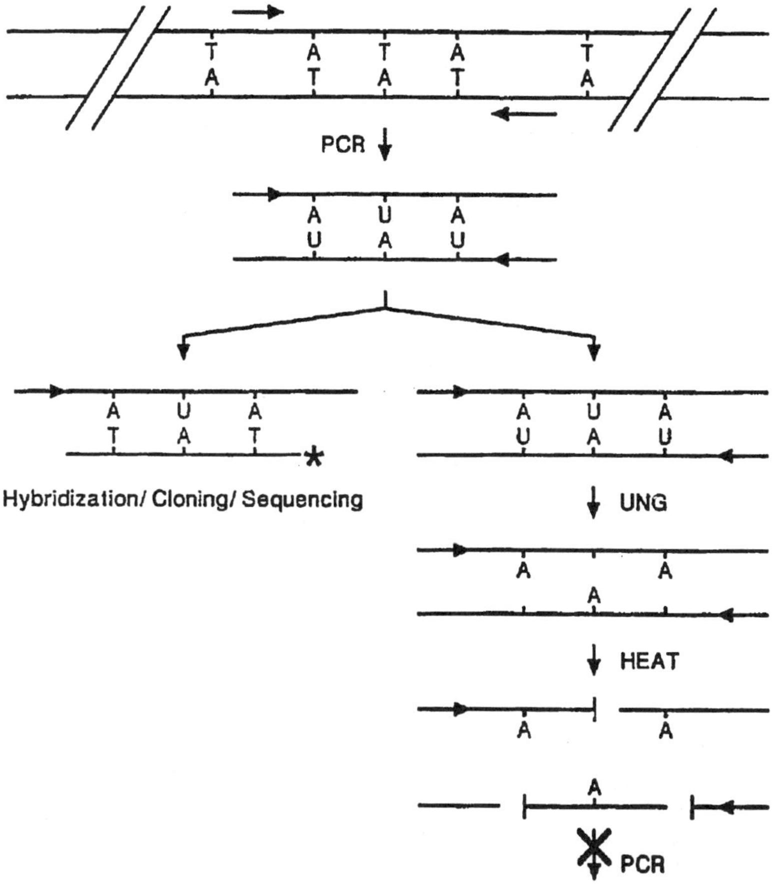

Figure 5.
Use of dUTP and uracil DNA glycosylase to eliminate errant PCR products as templates for subsequent amplifications leading to false positives.

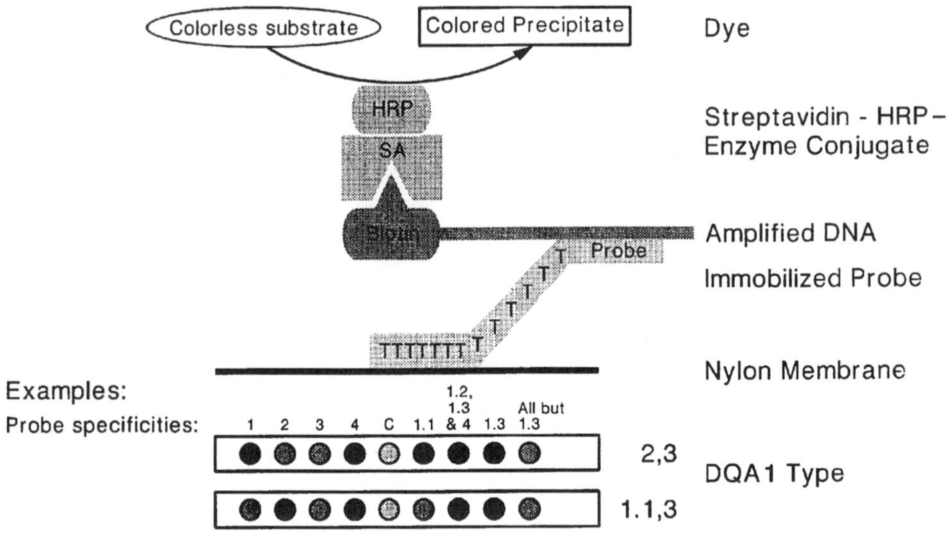

Figure 6.
Reverse dot blot hybridization to rapidly and precisely type HLA DQ alpha class II polymorphism. Schematic diagram of immobilized oligonucleotide probe detection of amplified DNA. A sequence-specific probe is "tailed" with a dT homopolymer and immobilized on a solid support. The amplified PCR product, which has incorporated a biotinylated primer, hybridizes to the probe. After washing away unbound DNA and primers, the biotinylated, amplified DNA binds an avidin horseradish peroxidase conjugate. The enzyme then converts a colorless dye into a colored precipitate. The format for detecting specific alleles in samples of amplified DQ alpha DNA, from heterozygous individuals is shown.

seropositive and culture positive samples was completed. The five laboratories participating in the study had significant experience with PCR, however the procedure and interpretive criteria varied somewhat and only one of the multiple primer pairs was used in common. The average sensitivity for the laboratories was 99.0 percent; the average specificity was 94.7 percent. One laboratory achieved 100 percent sensitivity and specificity. The overall false positive, false negative and indeterminate rates were 1.8, 0.8, and 1.9 percent, respectively.

This study demonstrated that PCR is a highly sensitive and specific assay for HIV–1 proviral DNA but that rigorous procedural and critical testing algorithms are required. Further, the two primer pair systems targeted to the gag gene showed I00 percent sensitivity and specificity. This observation suggests that the inability to detect all samples known to contain HIV–1 proviral DNA at the requisite level with different primer pairs may be due to a laboratory's experimental performance rather than a viral sequence variant incapable of amplification .

The lack of concordance between duplicate samples may be do to sample mixup. Resolution of sample mixup when using serological assays is difficult. Often, the sample is either rerun or another sample is taken for analysis. However, PCR assays for HIV, particularly if amplification of the histocompatibility region is used as a control for the number of cells examined and amplification integrity of the sample, allows for simple resolution. The pioneering PCR studies of Erlich and colleagues (Saiki *et al.*, 1986) on HLA genotyping were later exploited for HIV (Farzadeagan *et al.*, 1988) to demonstrate that due to the polymorphic nature of the region between the HLA DQ alpha DNA primers used, samples from different individuals could be discerned because of the differential hybridization of HLA sequence–specific probes (Saiki *et al.*, 1989) (see Figure 6) .

XIII. CARRY OVER

As mentioned above, considerable care must be taken to avoid carryover of DNA from one tube to another to prevent false–positives (Higuchi and Kwok, 1989). Because amplified sequences are present in large numbers, carryover of minute quantities of amplified DNA can lead to significant false–positive problems. The following is a list of procedures that should minimize carryover:

A. *Pre–and Post–PCR Reactions–Physical Separation:* To prevent carryover a separate room or containment unit, such as a biosafety cabinet, should be used for setting up amplification reactions. A separate set of supplies and pipettes should be kept in this area and should be used only for setting up PCR reactions. Care must be taken to ensure that amplified DNA is not brought into this area. Reagents, devices and supplies should never be taken and returned from an area where PCR analyses are being performed.

B. *Reagent Aliquotting:* Reagents should be aliquotted to minimize the number of repeated samplings. All reagents used in the PCR process must be prepared, aliquotted and stored in an area that is free of PCR–amplified product. Similarly, oligonucleotides used for amplification should synthesized and purified in an environment free of PCR product.

C. *Modified Pipettes:* Contamination of pipetting devices can result in cross–contamination of samples. To eliminate cross contamination of samples by pipetting devices, positive displacement pipettes with disposable tips or disposable tips with filters are recommended .

D. *Laboratory Technique:* The following precautions should be taken in all aspects of PCR from sample collection to PCR: (a) change gloves frequently, (b) uncap tubes carefully to prevent aerosols, (c) minimize sample handling, and (d) add non–sample components (mineral oil, dNTPs, primers, buffer and enzyme) to the reaction mixture before addition of sample DNA. Cap each tube after the addition of DNA before proceeding to the next sample.

E. *Selection of Controls:* For a positive control, select a sample that amplifies weakly but consistently. The use of strong positives will result in the unnecessary generation of large amounts of amplified DNA sequences. Well–characterized negative controls should also be used. The extreme sensitivity of the PCR process has the potential to amplify a nucleic acid sequence in a sample that is negative by all other criteria. Finally, multiple reagent controls should be included with each amplification. This is because the presence of a small number of molecules of PCR product in the reagents may lead to sporadic positive results. The reagent controls should contain all of the necessary components for the PCR process minus the template DNA.

CONCLUSION

As the AIDS epidemic continues to grow and spread, there will be an increasing need for sensitive and quantitative assays for HIV.

Quantitative tests for monitoring the infected cell load and cell–free virus in infected persons will be needed to monitor the clinical status of the patient and to evaluate the efficacy of new antiviral agents and potential vaccines.

The application of PCR technology to AIDS research opens up exciting new possibilities for the sensitive and specific direct detection and quantitation of HIV. PCR in its short existence has proven valuable for (a) detecting infection in seronegative persons, (b) quantifying the virus burden in a patient, (c) typing HIV infections, (d) measuring virus expression, (e) early diagnosis of perinatal transmission of HIV, and (f) resolving indeterminate Western blots. As a research tool the PCR technique is also proving useful for studying variant HIVs, distinguishing the important human retroviruses HTLV–I and HTLV–II and discovering new pathogenic human retroviruses. Further simplification of PCR technology with the addition of sensitive nonisotopic detection systems requiring less than 1 hour for a quantitative readout should guarantee PCR a significant role in the diagnosis of AIDS and HIV infection.

REFERENCES

Abrams, E. A., Rogers, M. F. and Ou, C–Y. (1990). *Proc. Fifth Intl. Conf. AIDS.* San Francisco.

Barre–Sinoussi, F., Chermann, J. C., Rey, F., Nugeyre, M. T. and Charmaret, S.m (1983). *Science* **220**, 868–871.

Becker–Andre, M. and Hahlbrock, K. (1989). *Nucleic Acid Res.* **17**, 9437–9446.

Blanche, S., Rouzioux, C. and Moscato, M. G. (1989). *New Engl. J. Med.* **320**, 1643–1648.

Boucher, C. A. B., Tersmette, M., Lange, J. M. A., Kellam, P., de Goede, R. E. Y. and Mulder, J. (1990). *Lancet* **336**, 585–590.

Byrne, B. C., Li, J. J., Sninsky, J. and Poiesz, B. J. (1988). *Nucleic Acids Res.* **16**, 4165.

Chadwick, E. G., Yogev, R., Kwok, S., Sninsky, J. J. Kellogg, D. E. and Wolinsky, S. M. (1989). *J. Infect. Dis.* **160**, 954–959.

Clavel, F. (1987). *AIDS* **1**, 135–140.

Clavel, F., Fuetard, D., Brun–Vezinet, Chamaret, S. and Rey, M. (1986). *Science* **233**, 343–346.

Clouse, K. A., Powell, D., Washington, I., Poll, G., Strebel, K. and Farrar, W. (1989). *J. Immunol.* **142**, 431–438.

Coombs, R. W., Collier, A. C., Allain, J–P., Nikora, B., Leuther, M., Gjerset, G. G. and Corey, L. (1989). *New Engl. J. Med.* **321**, 1626–1631.

Farzadeagan, H., Polis, M. A. and Wolinsky, S. M. (1988). *Ann. Intl. Med.* **108**, 785–790.

Foucault, C., Lopez, O., Jourdan, G., Fournel, J. J., Perret, P. and Gluckman, J. C. (1987). *Lancet* **i**, 165–166.

Gallo,k R. C., Salahuddin, S. Z., Popovic, M., Shearer, G. M. and Kaplan, M. (1984). *Science* **224**, 500–202.

George, J. R., Rayfield, M. A., Phillips, S., Heyward, W. L., Krebs, J. W., Odehouri, K., Soudre, R., DeCock, K. and Schochetman, G. (1989). *AIDS* **4**, 321–326.

Gilliland, G., Perrin, S., Blanchard, K. and Bunn, H. F. (1990). *Proc. Natl. Acad. Sci. USA.* **87**, 2725–2729.

Goodenow, M., Huet, T., Sasurin, W., Kwok, S., Sninsky, J. and Wain–Hobson, S. (1989). *J. Acquired Imm. Def. Synd.* **2**, 344–352.

Goudsmit, J., Paul, D. A., Lange, J., Speelman, H., Noordaa, J., VanDer Helm, H., deWolf, F., Epstein, L., Krone, W., Waolters, E., Loeske, J. and Coutinho, R. (1986). *Lancet* **ii**, 177–180.

Harper, M. H., Marselle, L. M., Gallo, R. C. and Wong–Stall, F. (1986). *Proc. Natl. Acad. Sci. USA* **83**, 772–776.

Hart, C., Chang, S–Y., Kwok, S., Sninsky, J., Ou, C–Y and Schochetman, G. (1990). *Nucleic Acid Res.* **18**, 4029–4030.

Hart, C., Spira, T., Moore, J., Sninsky, J., Schochetman, G., Lifson, A., Galphin, J. and Ou, C–Y. (1988). *Lancet* **ii**, 596–599.

Hewlett, I. K., Gregg, R. A. and Mayner, R. E. (1988). *Proc. Intl. Conf. AIDS.* **1**, 137.

Higuchi, R. and Kwok, S. (1989). *Nature* **339**, 237–238.

Ho, D. D., Moudgil, T. and Alam, M. (1989). *New Engl. J. Med.* **321**, 1625–1626.

Horsburgh, C. R. and Holmberg, S. (1988). *Transfusion* **28**, 192–195.

Horsburgh, C. R., Ou, C–Y., Jason, J., Holmberg, S. D., Longini, I. M., Schable, C., Mayer, K. H., Lifson, A. R., Schochetman, G., Ward, J. W., Rutherford, G. W., Evatt, B. L., Seage, G. R. and Jaffe, H. W. (1989). *Lancet* **ii**, 637–640.

Imagawa, D. T., Lee, M. H. and Wolinsky, S. M. (1989). *New Engl. J. Med.* **320**, 1458–1462.

Jackson, J. B., Kwok, S. Y. Sninsky, J. J. and Hopsicker, J. S. (1990). *J. Clin. Microbiol.* **28**, 16–19.

Jackson, J. B., MacDonald, K. L., Cadwell, J., Sullivan, C. Kline, W. E., Hanson, M. and Sannerud, K. J. (1990). *New Engl. J. Med.* **322**, 217–222.

Kellogg, D. E. and Kwok, S. (1989). In *PCR Protocols* (M. A. Innis, Gelfand, D. H., Sninsky, J. J. and White, T. J., eds.), pp. 337–347. Academic Press, New York.

Kellogg, D. E., Sninsky, J. J. and Kwok, S. (1990). *Anal. Biochem.* **189**, 202–208.

Kessler, H., Blaauw, B., Spear, J., Paul, D., Falk, L. and Landay, A. (1987). *J. Am. Med. Assoc.* **258**, 1196–1199.

Krivine, A., Vakudima, A. and LeMay, M. L. (1990). *J. Pediatrics* **116**, 372–376.

Kwok, S., Kellogg, D. E., McKinney, N., Spasic, D., Goda, L., Levenson, C. and Sninsky, J. (1990). *Nucleic Acid Res.* **18**, 999–1005.

Kwok, S., Mack, D. H., Mullis, K. B. Poiesz, B. and Erlich, G. (1987). *J. Virol.* **61**, 1690–1694.

Kwok, S., Mack, D. H., Sninsky, J., Erlich, G. D. Poiesz, B. J. and Dock, N. L. (1989). In *HIV detection by genetic engineering methods.* (P. A. Luciw and K. S. Steimer, eds.), pp. 243–255. Marcel Dekker, New York.

Larder, B. A. and Kemp, S. D. (1989). *Science* **246**, 1155–1158.

Larder, B. A., Darby, G. and Richman, D. D. (1989). *Science* **246**, 1731–1734.

Lion, T., Nighet, R., Hutchinson, M. A., Golomb, H. M. and Brownstein, B. H. (1989). *DNA* **8**, 361–367.

Loche, M. and Mach, B. (1988). *Lancet* **ii**, 418–421.

Longo, M. C., Berninger, M. S. and Hartley, J. L. (1990). *Gene* **93**, 125–128.

Meyerhans, A., Cheynier, R., Albert, J., Beth, M. and Kwok, S. (1989). *Cell* **58**, 901–910.

Mullis, K. B. and Faloona, F. A. (1987). *Methods Enzymol*. **155**, 335–350.

Murakawa, G. J., Zaia, J. A., Spallone, P. A., Stephens, D. A. and Kaplan, B. E. (1988). *DNA* **7**, 287–295.

Ou, C–Y. and Schochetman, G. (1989). In *Current communications in molecular biology. Polymerase chain reaction.* (H. A. Erlich, Gibbs, R. and H. H. Kazazian, Jr., eds.), pp. 165–170. Cold Spring Harbor Press, New York.

Ou, C–Y., Kwok, S., Mitchell, S. W., Mack, D. H., Sninsky, J. J., Feorino, P., Warfield, D., and Schochetman, G. (1988). *Science* **239**, 295–297.

Ou, C–Y., McDonough, S. H., Cabanas, D., Ryder, T. B., Harper, M., Moore, J. and Schochetman, G. (1990). *AIDS Res. Human Retroviruses* **6**, 1323–1329.

Pizzo, P. A., Eddy, J. and Falloon, J. (1988). *New Engl. J. Med.* **319**, 889–896.

Popovic, M., Sarngadharan, M. G., Read, E. and Gallo, R. C. (1984). *Science* **224**, 497–500.

Ranki, A., Valle, S. and Krohn, M. (1987). *Lancet* **ii**, 589–593.

Ratner, L. (1989). *AIDS Res. Human Retroviruses* **5**, 115–119.
Rayfield, M., DeCock, K., Heyward, W. L., Godstein, L. and Krebs, J. (1988). *J. Inf. Dis.* **158**, 170–176.
Rey, F. Salaun, D., Lesbordes, J. L., Gadelle, S. and Ollivier–Henry, F. (1986). *Lancet* **ii**, 1391–1392.
Rey, M. A., Girard, P. M., Harzic, M., Madjar, J. J., Brun–Vezinet, F. and Saimot, A. G. (1987). *Lancet* **i**, 388–389.
Rogers, M. F., Ou, C–Y., Kilbourne, B. and Schochetman, G. (1990). In *Pediatric AIDS, The challenge of HIV infection in infants, children and adolescents.* (P. A. Pizzo and C. M. Wilfert, eds.), pp. 159–174. Williams and Wilkins, Baltimore.
Rogers, M. F., Ou, C–Y., Rayfield, M., Thomas, P. A., Schoenbaum, E. E., Abrams, E., Krasinski, K., Selwyn, P. A., Moore, J., Kaul, A., Grimm, K. T., Bamji, M. and Schochetman, G. (1989). *New Engl. J. Med.* **320**, 1649–1654.
Saiki, R. K., Bugawan, T. L., Horne, G. T., Mullis, K. B. and Erlich, H. A. (1986). *Nature* **324**, 163–165.
Saiki, R. K., Gelfand, D. H., Stoffel, S., Scharf, S. J. Higuchi, R., Horn, G. T., Mullis, K. B. and Erlich, H. A. (1988). *Science* **239**, 487–491.
Saiki, R. K., Scharf, S., Faloona, F., Mullis, K. B., Horn, G. T., Erlich, H. A. and Arnheim, N. (1985). *Science* **230**, 1350–1354.
Saiki, R., Walsh, R. Levenson, C. H. and Erlich, H. A. (1989). *Proc. Natl. Acad. Sci. USA.* **86**, 6230–6234.
Schnittman, S. M., Greenhouse, J. J. Miltiades, B. S., Psallidopoulos, C., Baseler, M. and Salzman, N. P. (1990). *Ann. Intl. Med.* **113**, 438–443.
Schnittman, S. M., Psallidopoulos, M. C., Lane, H. C. Thompson, L., Baseler, M. and Massari, F. (1990). *Science* **245**, 305–308.
Scott, G. B., Fischl, M. A., Klimas, N., Fletcher, M. A., Dickinson, G. M., Levine, R. S. and Parks, W. P. (1985). *J. Am. Med. Assoc.* **253**, 363–366.
Shaw, G. M., Hahn, B. H. Arya, S. K. Groopman, J. E. and Gallo, R. C. (1984). *Science* **226**, 1165–1171.
Simmonds, P., Balfe, P., Peutherer, J. F., Ludlam, C. A., Bishop, J. O. and Leigh–Brown, A. J. (1990). *J. Virol.* **64**, 864–872.
Simmonds, P., Lainson, F. A., Cuthbert, R., Steel, C. M. Peutherer, J. F. and Ludlam, C. A. (1987). *Brit. Med. J.* **296**, 593–598.
Spear, G. T., Ou, C–Y., Kessler, H. A., Moore, J. L. Schochetman, G. and Landay, A. L. (1990). *J. Inf. Dis.* **162**, 1239–1244.
Wang, A. M., Doyle, M. V. and Mark, D. F. (1989). *Proc. Natl. Acad. Sci. USA.* **86**, 9717–9721.
Ward, J. W., Schable, C., Dickinson, G. M., Makowka, O. L., Yanaga, G. A., Caruana, R., Chan, H., Salazar, F., Schochetman, G. and Holmberg, S. (1989). *Transplantation* **47**, 722–724.
WHO (1993). *Weekly Epidemiol. Rec.* **68**, 193–200.
Wolinsky, S. M., Rinaldo, C. R., Kwok, S., Sninsky, J. J., Gupta, P., Imagawa, D., Farzadegan, H., Jacobson, L. P., Grovit, K. S., Lee, M. H., Chmiel, J. S., Ginzburg, H., Kaslow, R. A. and Phair, J. P. (1989). *Ann. Intl. Med.* **111**, 961–972.

3

The Use of Transcription–based Amplification Systems in the Diagnosis of HIV–1 Infections

Deborah Y. Kwoh and Thomas R. Gingeras

I. INTRODUCTION

A. Nucleic acid based hybridization for detection of viruses

The advances in nucleic acid sequencing and detection by hybridization have opened new approaches to the diagnosis of human pathogens. While nucleic acid hybridization made detection of target sequences highly specific and/or relatively rapid in comparison to standardized immunological or culture–based assays, the methodology was slow in being adapted to clinical settings for two major reasons. First, the most sensitive nucleic acid hybridization techniques requires the use of radiolabeled probes making everyday utilization and waste management problematic. Secondly, even using radioactivity, nucleic acid hybridization is unable to detect targets below approximately 10^4 molecules (Chou and Merigan, 1982). This limitation in sensitivity poses a major stumbling block for the use of probes especially in viral diagnostics since several clinically important viruses such as human T–cell lymphoma/leukemia virus type 1 and 2 (HTLV–1 and 2), cytomegalovirus (CMV) and especially human immuno–deficiency virus type 1 (HIV–1) frequently are present clinical samples well below the level of sensitivity achievable by most nucleic acid–based hybridization assays (Landry, 1990). A major breakthrough in the utility of nucleic acid hybridization as a diagnostic tool occurred with the introduction of *in vitro* nucleic acid target amplification. The development of the polymerase chain reaction (PCR) in 1985 (Saiki *et al.*, 1985) permitted the amplification of target nucleic acid sequences over a million–fold, thereby allowing a single copy

Deborah Y. Kwoh and Thomas R. Gingeras • Baxter Diagnostics, Inc., Life Sciences Research Laboratory, P.O. Box 910492, San Diego, CA 92191–0492.

Applied Virology Research, Vol. 3, Edited by
E. Kurstak *et al.*, Plenum Press, New York, 1994

of a target molecule to be detected in a clinical sample. The only other diagnostic tool comparable in sensitivity is the *in vitro* propagation by culturing of the pathogen.

B. Methods of nucleic acid amplification based on DNA replication and ligation and RNA transcription

The PCR amplification system was conceived by Dr. K. Mullis and developed by co–workers (Mullis, 1990). This amplification method relies on repeated cycles of DNA replication catalyzed by DNA polymerase–mediated primer extension (Saiki *et al.*, 1985; Mullis and Faloona, 1987; Saiki *et al.*, 1988). Each cycle of replication is preceded by periods of thermal denaturation (Figure 1A). Typically, greater than twenty cycles of PCR are required to produce better than 10^6–fold amplification of a target sequence. The PCR protocol has been adapted for a wide variety of applications revealing the considerable potential of target amplification/nucleic acid–based assays (Erlich *et al.*, 1988; Erlich, 1989; Persing and Landry, 1989). PCR has been used to detect clinically important viruses such as HTLV–1 and 2 (Abbott *et al.*, 1988; Bhagavati *et al.*, 1988; Buchbinder *et al.*, 1988; Duggan *et al.*, 1988; Kwok *et al.*, 1988), CMV (Demmler *et al.*, 1988; Shibata *et al.*, 1988b), hepatitis B (Larzul *et al.*, 1988; Zeldis *et al.*, 1989; Paterlini *et al.*, 1990a), human papillomavirus (Shibata *et al.*, 1988a; Shibata *et al.*, 1988c), rhinovirus (Gama *et al.*, 1988), and hepatitis delta virus (Zignego *et al.*, 1990). Importantly, PCR amplification has found extensive application for detecting human immunodeficiency virus (HIV–1). The technique has been used for the detection of HIV–1 proviral sequences in infected individuals prior to seroconversion (Imagawa *et al.*, 1989; Pezzella *et al.*, 1989; Wolinsky *et al.*, 1989; Hewlett *et al.*, 1990), for confirmation of an HIV–1 diagnosis in seropositive adults (Farzadegan *et al.*, 1988; Hart *et al.*, 1988; Murakawa *et al.*, 1988; Ou *et al.*, 1988; Rayfield *et al.*, 1988; Jackson *et al.*, 1990; Lifson *et al.*, 1990; Sheppard *et al.*, 1991) and for screening neonates of HIV–1 infected mothers (Laure *et al.*, 1988; Rogers *et al.*, 1989; Paterlini *et al.*, 1990b; Williams *et al.*, 1990; de la Salle *et al.*, 1991).

Since the first report of the PCR *in vitro* target amplification system, other nucleic acid target amplification methods have been described, exploiting the properties of different enzymatic reactions. DNA ligase has been used to amplify target nucleic acid sequences by catalyzing the covalent attachment of oligonucleotides which hybridize to adjacent sequences in the target genome (Figure 1B). This ligase–based amplification has various appellations: ligation amplification reaction (LAR) (Wu and Wallace, 1989), ligase–based amplification system (LAS) (Barringer *et al.*, 1990) or ligase chain reaction (LCR) (Barany, 1991a; Barany, 1991b) which use phage T4, *Escherichi coli* or cloned thermostable ligases, respectively. A target amplification system based on RNA replication using the Qβ–replicase has also been described (Lizardi *et al.*, 1988; Kwoh and Kwoh, 1990).

Two amplification systems have been reported which use RNA transcription as the principal means of increasing the copy number of the target nucleic acid. The transcription–based amplification system (TAS) (Kwoh *et al.*, 1989) employs multiple two–step cycles to amplify target sequences (Figure 2A). The first step of each cycle is the synthesis of a double–stranded cDNA which requires thermal denaturation for synthesis of the second strand of cDNA. A T7 promoter is incorporated into the cDNA via one of the primers used in the synthesis reaction. The second step employs T7 RNA polymerase to transcribe the promoter containing cDNA template, producing 10–100 transcripts per cDNA template. The TAS reaction yields better than 10^6–fold amplification in 4 cycles (Kwoh *et al.*, 1989) and the results of it application in detecting HIV–1 in infected peripheral blood mononuclear cells (PBMCs) were shown to be comparable to PCR amplification of the same samples (Davis *et al.*, 1990).

The second transcription–based amplification system evolved from the TAS protocol and functions in an isothermal and self–cycling manner. This modified reaction is called self–sustained sequence replication (3SR) (Guatelli *et al.*, 1990). A schematic drawing of the self–sustained sequence replication (3SR) reaction is shown in Figure 2B. An RNA target is annealed to a primer containing the T7 RNA polymerase binding sequence (promoter) upstream of the 5' end of a target complementary sequence (TCS). Reverse transcriptase synthesizes a first strand cDNA using the annealed oligonucleotide as the primer. RNase H can then cleave the RNA in the RNA:cDNA heteroduplex, releasing the first strand cDNA for hybridization to a second oligonucleotide primer (B) and extension of this second oligonucleotide by reverse transcriptase. Completion of the second strand cDNA synthesis produces a double–stranded cDNA with a functional T7 RNA polymerase promoter at the 5' end. T7 RNA polymerase can utilize this cDNA template to synthesize an RNA product which is complementary (antisense) to the original RNA target. This RNA product serves as a template for further amplification through synthesis of the cDNA intermediate, primed first by oligonucleotide B and then by oligonucleotide A, and further accumulation of the RNA product.

Optimal conditions have been determined in which all three of the required enzymes [Avian Myeloblastosis Virus reverse transcriptase (AMV RT), *E. coli* RNase H and T7 RNA polymerase] can function in concert (Fahy *et al.*, 1991). The optimized 3SR protocol is shown in Figure 3. Under certain 3SR reaction conditions, the RNase H activity of *E. coli* can be replaced by the RNase H activity of the AMV RT. This requires supplementing the 3SR reaction buffer with 10%(v/v) dimethylsulfoxide (DMSO) and 15%(w/v) sorbitol (Fahy *et al.*, 1991).

DNA targets can be amplified as efficiently as RNA targets, but require certain additional steps at the beginning of the 3SR reaction. Target molecules must first be denatured prior to hybridization with primer A and reverse transcription step using AMV RT. In a second step, the target:cDNA duplex is thermally denatured, and annealed to primer B prior to the 3SR reaction protocol described in Figure 3.

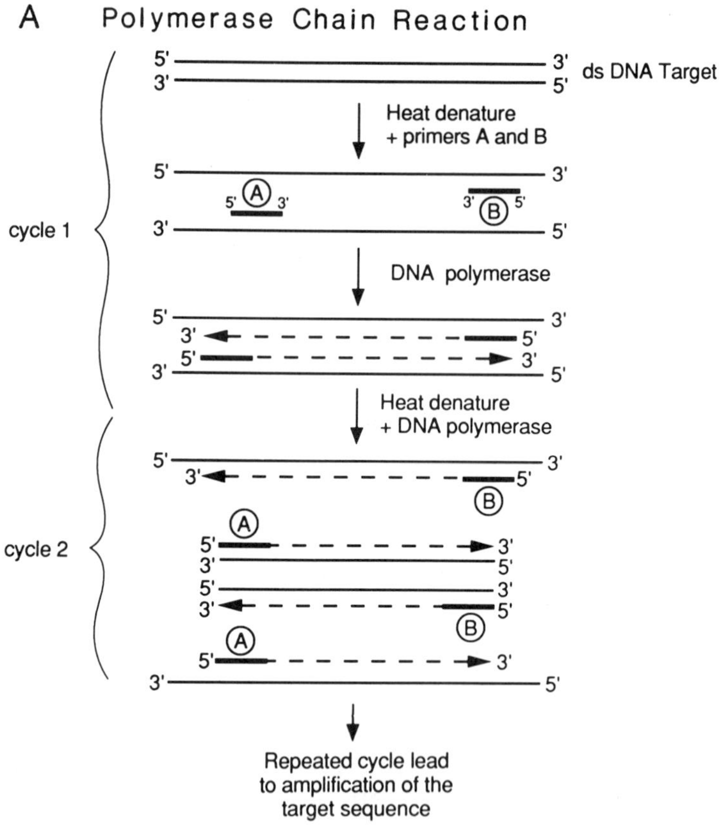

Figure 1.
Panel A: A diagram of the steps required for amplification of a DNA sequence by the polymerase chain reaction (PCR). Two oligonucleotide primers are complementary to sequences located several hundred base pairs apart on opposite strands of the target. A cycle of PCR consists of thermal denaturation of the target in the presence of a large excess of primers, followed by primer annealing and extension. The product molecules can serve as templates in the following cycles. Repeated cycles lead to exponential accumulation of a portion of the target region defined by the two primers.

B. Ligase-based Amplification

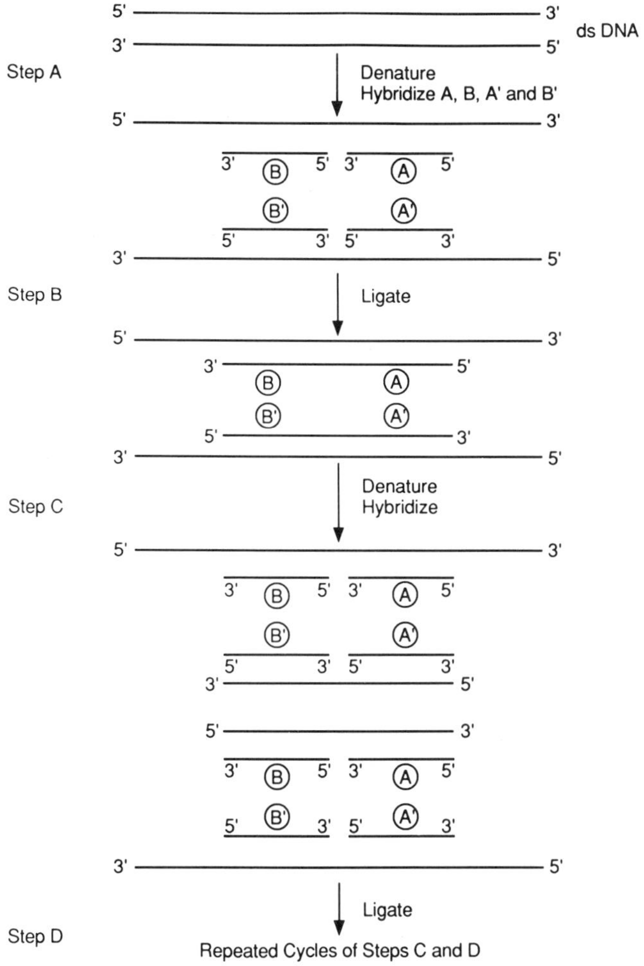

Panel B: A schematic diagram of a liagase–based amplification system. A double–stranded DNA target is thermally denatured i the presence of two sets of complementary oligonucleotdies A/A′ and B/B′ where oligonucleotides A and B are complementary to adjacent sequences on the target DNA. The oligonucleotides A/B and A′/B′ hybridize to opposite strands of the denatured DNA. A DNA ligase joins the 5′–ends of oligonucleotides B and A′ to the 3′–ends of oligonucleotides A and B′, respectively (step B). the ligated oligonucleotids are then thermally denatured from the target DNA. Oligonucleotides A/B and A′/B′ are again hybridized to the appropriate target strands and to the ligated product strands (step C). These oligonucleotides are then joined by a DNA ligase (step D). Steps C and D are repeated to achieve exponential amplification of the ligated products AB and A′B′.

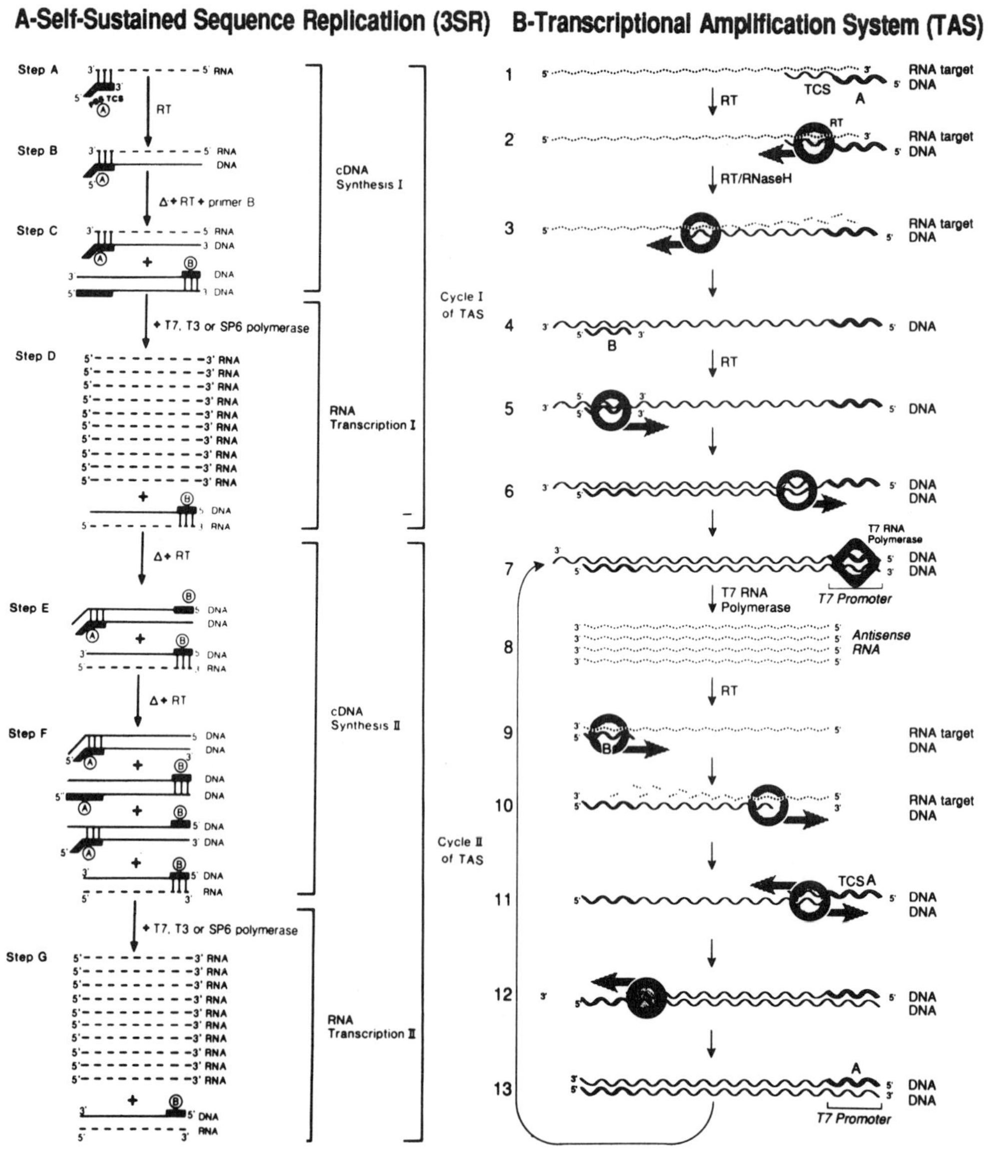
A-Self-Sustained Sequence Replication (3SR)
B-Transcriptional Amplification System (TAS)
Step A
Step B
Step C
Step D
Step E
Step F
Step G
RT
Δ + RT + primer B
+ T7, T3 or SP6 polymerase
Δ + RT
RNA
DNA
cDNA Synthesis I
RNA Transcription I
cDNA Synthesis II
RNA Transcription II
Cycle I of TAS
Cycle II of TAS
RNA target
RT/RNaseH
T7 RNA Polymerase
T7 Promoter
Antisense RNA
TCS
TCS A

Figure 2.
Panel A: A schematic diagram of the two-cycle transcription-based amplification system (TAS) reaction for amplifying an RNA target. The target RNA (—) is hybridized to a primer oligonucleotide (primer A) that contains a promoter binding site (PBS) for T7 (T3 or SP6) RNA polymerase and a target-complementary sequence (TCS) (step A). RT elongates primer A to yield a newly sysnthesized DNA strand complementary to the target RNA (step B). The RNA:DNA heteroduplex is denatrued by heat (Δ) and primer oligonucleotide B is annealed to the newly synthesized DNA strand containing the PBS. RT is added to produce a double-stranded cDNA and a new DNa:RNA heteroduplex (step c). Incubation of the double-stranded cDNA with T7 (T3 or SP6) RNA polymerase results in the synthesis of multiple RNA transcripts form the PBS-containing double-stranded DNA template (step D). Some of this RNA is immediately converted to RNA:DNA heteroduplex by RT (still in the reaction mixture form step C) using oligonucleotide B as a primer. Further amplification of target sequences can be obtained by a second cycle of cDNA synthesis (steps E and F) and RNA transcription (step G).

Panel B: Strategy of the 3SR amplification scheme. The 3SR reaction consists of continuous cycles of reverse transcription and RNA transcription designed to relicate a nucleic acid (RNA-target in figure) using a double-stranded cDNA intermediate. Oligonucleotides A (containing a non-homologous T7 promoter sequence) and B primer DNA synthesis sproducing a double-stranded cDNA containing a functional double stranded T7 promoter (steps 1–6). Complete cDNA synthesis is dependent upon digestion of RNA in the intermediate RNA:DNA hybrid by RNase H (step 3). Transcription competent cDNAs are used to produce multiple (50–1000) copies of antisense RNA transcript of the original target (steps 7–8). these antisense RNa transcripts are immediately converted to T7-promoter containing double-stranded cDNA copies (steps 9–12) and used again as transcription templats. This process continues in a self-sustained cyclic fashion under isothermal conditions (42°C) until components in the reaction become limiting or inactivated (enzymes). Dotted lines, RNA; thin lines DNA; thick lines, T7 promoter sequence (5′-AATTTAATACGAC-TCACTATAGGGA-3′; Fahy *et al.*, 1991); circles, reverse transcriptase; diamonds, T7 RNA polymerase; TCS, target complementary sequence.

C. Distinguishing features of the 3SR amplification system

Some of the advantages and disadvantages of the *in vitro* target amplification systems mentioned above have been reviewed elsewhere (Persing and Landry, 1989; Gingeras *et al.*, 1990a; Kwoh and Kwoh, 1990; Gingeras and Kwoh, 1991; Persing, 1991). Interestingly, the 3SR amplification reaction has several features which distinguish it from other methods for nucleic acid amplification. The unique attributes of the 3SR method are its rate of amplification, and its self–cycling and isothermal characteristics. Depending on the target sequence being amplified, the 3SR reaction produces a 10^6–10^8–fold increase in target sequence in 1 hour at 42°C (Fahy *et al.*, 1991). Similar levels of amplification can be achieved using the PCR method but may require 3–4 hours and 30 reaction cycles involving 90 temperature changes. To minimize the number of manipulations in PCR, the use of automated thermocycling equipment is essential. In contrast, the 3SR method merely requires a constant temperature bath.

Arguably, the most distinguishing feature of the 3SR reaction is its substrate selectivity. Although the 3SR amplification reaction can amplify both double–stranded DNA and single–stranded RNA targets with equivalent efficiencies, (Gingeras *et al.*, 1990b), the DNA target amplification requires two thermal denaturation steps prior to 3SR amplification as described above. Thus, if a 3SR reaction is initiated at 42°C, only RNA molecules serve as target sequences for amplification. This substrate selectivity of the 3SR reaction, therefore, makes it uniquely suited for the selective amplification of RNA targets in the presence of DNA copies of the same sequence. This feature of the 3SR amplification reaction might lead to assays targeting mRNA to distinguish between productive and latent infections of cells by viruses with DNA or double–stranded RNA genomes, or between actively expressing and inactive cellular or proviral genes. In contrast, specific amplification of mRNA using PCR requires an extra reverse transcription step, and also requires the removal of DNA from the target sample prior to amplification. Another approach to distinguish mRNA target sequences from genome DNA copies by PCR is to select primers in adjacent exons and differentiate the amplification products of the RNA and DNA targets by the absence or presence of intron sequences. However, the latter approach can be compromised by the presence of pseudogenes or incompletely processed mRNAs.

The major product of the 3SR reaction is a single–stranded RNA. While the electrophoretic assays for the 3SR RNA product require denaturing conditions for accurate size measurements and thus are more cumbersome than for the DNA product of PCR amplifications, the single–stranded RNA of the 3SR reaction has two advantageous features. First, the RNA product of the 3SR reaction can be sequenced directly because of its single–stranded nature (Guatelli *et al.*, 1990; Gingeras *et al.*, 1991). In contrast, asymmetric PCR (Gyllensten and Erlich, 1988; Gibbs *et al.*, 1989; McCabe, 1990) requires additional amplification using unequal primer concentrations in order to produce a single–stranded product for sequencing (Wedell *et al.*, 1990). Second, the single–stranded 3SR product is amenable to rapid detection, genotypic characterization and quantitation using bead–based sandwich hybridization (BBSH) reactions.

In BBSH reactions, the 3SR product is simultaneously hybridized to an excess of labeled probe oligonucleotide and capture oligonucleotide, which is attached at its 5'–end to Trisacryl beads, (termed Oligobeads; Davis *et al.*, 1990; Fahy *et al.*, 1991). The percent of the label associated with the beads allows a direct calculation of the number of femtomoles of target detected by sandwich hybridization. The hybridization is quantitative when the product is within the linear range of the hybridization system. Using 50 femtomoles of labeled probe oligonucleotide (~2 x 10^5 cpm) and 25 mg of Oligobeads, the linear range of the BBSH is between 100 attomoles and approximately 25 femtomoles of product as shown in Figure 4A and B. Above 50 femtomoles of target, the probe becomes limiting under these conditions, while background signal obscures detection below 100 attomoles. As anticipated, the linear range of the BBSH assay is narrowed (100 attomoles to 10 femtomoles of product detected) using 20 femtomoles of probe. The variability of the BBSH assay is typically 5% or less for duplicate assays. In conjunction with 3SR amplification, the BBSH detection assay can easily detect 1000 copies of the HIV–1 target which is approximately the copy–number of HIV–1 in one infected cell (Richman *et al.*, 1987).

Quantitation methods for PCR products involves a) scintillation counting of directly or indirectly labeled PCR products after excision from agarose or acrylamide gels; b) densitometer readings of autoradiograms of directly labeled PCR products after fractionation by gel electrophoresis; or c) densitometer tracings after Southern blotting of unlabeled product from gels to membranes (or slot blots) followed by hybridization to labeled probes. The laborious nature and the variability pose significant drawbacks to these methods (Abbott *et al.*, 1988).

II. APPLICATIONS OF 3SR AMPLIFICATION REACTIONS

A. Detection of AZT–resistant HIV–1 using 3SR amplification followed by differential bead–based sandwich hybridization BBSH

Development of effective chemotherapeutic regimens for HIV–1, based on inhibition of the HIV–1 polymerase, are likely to require the ability to monitor patients for the appearance of drug–resistant viruses (Richman, 1990). The diagnostic approach of isolating virus from patient samples and then testing the isolated virus for drug sensitivity (Larder *et al.*, 1990) is both time consuming and costly. Assays which directly detect virus isolates with mutations in the polymerase gene *(pol)* codons which confer drug–resistance are being developed by several groups (Boucher *et al.*, 1990; Gingeras *et al.*, 1991; Japour *et al.*, 1991; López–Galíndez *et al.*, 1991).

Prolonged administration of Zidovudine (AZT) to HIV–1 infected patients has been shown to result in the emergence of drug–resistant isolates of the virus (Larder *et al.*, 1989; Richman *et al.*, 1990). Changes in four codons (codon 67, 70, 215 and 219) within the *pol* gene have been identified as contributing to the AZT–resistant phenotype of HIV–1 (Larder and Kemp, 1989). The 3SR amplification/BBSH assay for detection of mutations in the *pol* gene of HIV–1 consists of amplifying the 67–70 and 215–219 amino acid codon–containing regions in separate 3SR reactions. The products of these reactions are then detected by differential BBSH using probes which specifically detect either wild–type or mutant sequences in the codons of interest (Gingeras *et al.*, 1991). In the differential BBSH procedure, the amounts of 3SR product hybridized to allele-specific detection probes are compared under low and high stringency conditions (Table 1). The genotype of the viral isolate is determined by the genotype of the detection probe which remains hybridized at the more stringent hybridization conditions (i.e., low salt and elevated temperatures). The specific detection of wild–type and mutant sequences at amino acid codons 67, 70, 215 and 219 was first demonstrated on 3SR amplification products of RNA and DNA targets with defined sequences. Interestingly, short oligonucleotide probes (15–20 nucleotides) designed to determine the genotype at each locus individually (67, 70, 215 or 219 codon position only) failed to hybridize efficiently to the 3SR products in the BBSH format presumably due to secondary structures formed by the RNA products. Longer oligonucleotides (27–30 nucleotides) proved to be more successful in the differential BBSH assay, because they provide greater stability of the RNA:DNA duplex. However, because of the length of these oligonucleotides and the proximity of the mutation sites, mutations at two positions (codons 67 and 70 or codons 215 and 219) were monitored simultaneously.

The 3SR amplification/differential BBSH assay was applied to HIV–1 which had been isolated after co–cultivation of uninfected MT–2 cells with PBMCs from infected individuals receiving AZT therapy. The genotype for each locus (67, 70, 215 and 219) of the viral isolates was determined and compared to the results obtained on the same samples assayed by a PCR/Southern blot analysis of the same isolates (or analysis directly on PBMCs of the same patient samples) and to the nucleotide sequence obtained directly from the 3SR amplified product. This comparison led to several important conclusions:

1. The sequence data and the differential BBSH assay were in agreement except for the 67 and 70 codons of two samples where the sequence data suggested a mixed population with the mutant as the predominant form, while BBSH detected only the mutant form.
2. The discrepancies between the PCR/Southern hybridization assay and the 3SR/BBSH results could be explained by mutations at sites unrelated but proximal to the particular mutations of interest, resulting in poor hybridization of the 3SR product to the capture Oligobead.
3. The viral genotype obtained after propagation of the HIV–1–infected samples in the co–culture system differed from the viral genotype detected

Table 1. Genotype Determination of Alleles of AZT Resistants by 3SR/Differential BBSH Assay

A. Amino Acid Regions 67–70[1]

Target	Hybridization Gen. String.	Femtomoles Product Detected with Probes[2] 90–36 wt/wt	90–48 mut/mut	90–49 wt/mut	90–50 mut/wt	89–441 control
pPol 18[3]	Low	26.1	2.1	27.9	11.7	11.0
	High	4.3	0.3	0.1	0.2	3.5
pHIVRTMC[4]	Low	1.8	42.7	45.7	21.3	33.8
	High	—[5]	5.2	0.7	—	13.7
pPol 67–18[6]	Low	19.8	11.8	2.2	20.9	15.0
	High	0.2	0.4	0.4	6.8	5.3
pPol 70–18[7]	Low	17.6	25.4	42.1	—	18.6
	High	—	0.6	18.8	0.1	9.0

B. Amino Acid Regions 215–219[8]

Target	Hybridization Gen. String.	Femtomoles Product Detected with Probes[2] 90–303 wt/wt	90–304 mut/mut	90–305 wt/mut	90–302 mut/wt	89–535 control
pARV[3]	Low	8.0	0.2	1.0	0.5	31.2
	High	6.1	—	0.6	—	27.4
pHIVRTMC[4]	Low	0.3	15.8	12.3	9.6	29.7
	High	—	14.9	—	0.1	34.5
pPol 215P–18[6]	Low	1.7	5.6	0.1	13.4	12.9
	High	—	0.4	0.2	2.7	18.9
pPol 215T–18[6]	Low	1.7	6.3	0.9	5.8	20.6
	High	—	0.2	—	1.7	19.1
pPol 219–18[7]	Low	1.0	3.4	13.3	0.1	16.3
	High	0.4	—	14.2	—	NT[8]

(continued)

Table 1 (continued)

---

1 RNA transcripts (0.1 attomole) were used as targets for 3SR amplification reactions according to the protocol described in the legend to Fig. 3. Transcription plasmids (1–10 μg) were lenearized with EcoRI or HindIII. T7–generated antisense transcripts were made using 1000 units of T7 enzyme. Quantitation of RNA product was performed by spectroscopic measurement. (Gingeras *et al.*, 1991b)

2 The sequence for the probes is given in Gengeras *et al.* (1991b), as well as the conditions used for the low and high stringency BBSH conditions.

3 wt/wt

4 mut/mut

5 less than 0.1 femtomoles detected

6 mut/wt

7 wt/mut

8 Not tested

---

directly by amplification of viral sequences in PBMC samples of the same patient.

4. Two samples had the same genotype of wild–type/mutant at codons 67/70 and mutant/wild–type at codons 215/219 but differed in their susceptibility to AZT suggesting that sequences at positions other than the four monitored positions play a role in the development of viral resistance to AZT.

An additional mutation at codon 211 which was identified by sequencing of the 3SR amplification products was present in 4 of the viral isolates which were resistant to AZT and might represent an additional sequence involved in conferring the AZT resistant phenotype (Gingeras *et al.*, 1991).

The 3SR amplification/differential bead–based assay has also been used to detect nevirapine–resistant HIV–1 isolates. Nevirapine (BI–RG–587) is a non–competitive inhibitor of HIV–1 reverse transcriptase which interacts with the highly conserved tyrosine residues at positions 181 and 188 of the enzyme (Richman *et al.*, 1991). Substitution of a cysteine residue for the wild–type tyrosine residue at amino acid position 181 within the HIV–1 reverse transcriptase is associated with viral resistance to nevirapine. The 3SR amplification/differential BBSH assay detected the emergence of nevirapine–resistance by following changes in the codon sequence of amino acid 181 in the reverse transcriptase of HIV–1 during several passages of the virus in the presence of the non–nucleoside inhibitor. The drug–resistant virus appeared as early as the second passage of HIV–1 in nevirapine. The resistant phenotype of these isolates remained stable with continued passage in the absence of nevirapine.

The differential BBSH protocol used for the detection of mutations involved in AZT resistance was modified for the detection of nevirapine resistant virus in this study. In the AZT study only two hybridization stringencies were employed to distinguish between targets which were complementary or non–complementary to a particular detection oligonucleotide. Complementary or non–complementary targets were distinguished in the nevirapine study by continuous monitoring of the amount of hybridized detection probe through a series of increasingly stringent washes. This procedural modification allows detection of targets which might have slightly altered hybridization characteristics due to base changes at sites within the sequences complementary to probe or bead capture oligonucleotide but which are not relevant to the mutations being monitored.

B. Development of a quantitative 3SR amplification assay for HIV–1

The use of the 3SR/BBSH assay for the detections of drug–resistant HIV–1 isolates illustrates the utility of the protocol as a qualitative assay. More demanding are those clinical or biological questions which require quantitative measurements. A quantitative assay is defined as a procedure which is capable of determining the amount of target in cells or the number of target molecules in a sample. The amount of target can be measured in absolute numbers or in relative numbers with respect to a reference target. Development of a quantitative diagnostic assay has become important for the detection of HIV–1 in clinical samples to 1) follow the viral load in an infected patient over the time course of the disease, 2) determine the increase or decrease in expression of particular viral genes (or host genes induced or repressed by the virus) over time during an infection and 3) follow viral load during drug therapy regimens to determine the efficacy of treatment. A complete assay for determining the amount of target nucleic acid in a biological sample requires that all steps in the assay, namely sample preparation, amplification, hybridization and detection, have adequate controls to ensure accuracy and reproducibility. Steps such as sample preparation to liberate the target nucleic acid and the amplification reaction itself by nature involve many variables, making quantitation a complex task.

1. Sample preparation

Sample preparation is the first crucial step in obtaining a reliable and quantitative amplification assay and is probably the area least studied. Ideally the procedure should be: 1) rapid and simple; 2) compatible with the amplification reaction; 3) generic (i.e., applicable to most clinical sample types),; and 4) reproducible. The fewer steps employed and the less hazardous (but still virus inactivating) the chemicals used during sample preparation, the more attractive the procedure is to clinical labs and the more applicable to automation.

The extraction procedure for the target must reproducibly liberate the nucleic acid in a usable form, i.e., sonication steps are unacceptable. Any impurities which are carried over from the sample preparation procedure, i.e., from the clinical sample itself or from the sample extraction buffer, must not interfere with the enzymes in the 3SR amplification reaction. A generic procedure which could be used for isolation, for example of HIV–1, from all sample types is desirable but difficult to achieve since HIV is found in quite a variety of sample types, e.g., blood, sputum and serum, each having different "impurities" which must be removed prior to amplification. To ensure fidelity of the sample preparation step and to determine target recovery, a control nucleic acid sequence should be included during sample preparation. To date only one published report describing a quantitative PCR test has controlled this step of the assay (Gaudette and Crain, 1991). This group added a specific amount of labeled RNA during the sample preparation to follow recovery at each step. For clinical samples, it is equally important to determine the effects of sample collection and storage on the recovery of target nucleic acids.

2. Amplification

Estimation of the initial target concentration after *in vitro* amplification of a sample requires that there should be a definable relationship between the amount of input target nucleic acid and the amplified product over the time of the amplification reaction. Because the 3SR amplification reaction relies on the concerted action of 3 enzymes, a variety of factors must be controlled to ensure that the amplification reaction proceeds in a manner in which this definable relationship is manifested. The kinetic rates of the 3SR reaction have been measured for target amounts ranging from 10^2 to 10^5 molecules (0.002 to 2.0 attomoles). These kinetics are characterized by a rapid rate of product accumulation for the first 10 minutes, during which the 3SR products increase 10– fold about every 2.5 minutes. This is followed by a slowing of the rate over the next 20 minutes such that at 30 minutes into the reaction, 3.75 minutes are needed to produce a 10–fold increase in product (Gingeras and Kwoh, 1991). Importantly, the rate profiles for each target concentration are parallel between the 20 and 30 minute time period, thereby providing a linear relationship between the amount of product detected and the initial target concentration (Figure 5). This observation is valid until the total 3SR product reaches approximately 50 femtomoles/µl. At later times (>30 minutes) the kinetic rate curves change as a function of target concentration. The lack of correlation between accumulated and input target when the product exceeds 50 femtomole/µl most likely reflects substrate (primer) depletion and possibly enzyme instability in the 3SR amplification reaction. Other factors which can influence the kinetics of the 3SR reaction are the sequences of the amplified target (Fahy *et al.*, 1991) and the 3SR primers (which can influence the amount of non–specific product synthesized). Therefore, the time interval in the 3SR reaction during which the direct relationship between input target and amplified product

a To an autoclaved 1.5 ml Eppendorf tube add:
20 µl 5X buffer (200 mM Tris HCl, pH 8.1, 150 mm $MgCl_2$, 100 mM KCl, 50 mM DTT, 20 mM spermidine)
5 µl each primer (0.1 µM each, final concentration)
4 µl dNTP mix (1 mM final)
24 µl HIV–1 RNA (0.1 attomoles)
37 µl H_2O (total volume = 100 µl)

Use H_2O instead of RNA for water control reactions.

b Denature at 65°C for 1 minute and transfer to 42°C.

c Add enzymes as a mix: 30 units AMV reverse transcriptase, 4 units RNase H and 100 units T7 RNA polymerase. In the case of the two–enzyme reaction, 10% (v/v) dimethylsulfoxide (DMSO) and 15% (w/v) sorbitol are included and the enzyme concentrations are 10 units AMV RT and 20 units T7 RNA polymerase. This mix can be stored on ice for up to 1 hours.

d After enzyme addition, mix by flicking the tube and incubate at 42°C for 1 hour.

e Stop the reaction by freezing the tube on dry–ice. Reactions may be stored at –20°C.

All reagents were prepared with 0.1% diethylpyrocarbonate (DEPC)-treated water to suppress nuclease activity. AMV reverse transcriptase was obtained from Life Sciences, Inc., *E. coli* RNase H was purchased from Bethesda Research Laboratories and T7 RNA polymerase was obtained from Stratagene. Oligonucleotides were synthesized by phosphoramidite chemistry using an Applied Biosystems 380 A synthesizer. The amount of enzyme required in the 3SR reaction might vary depending on the supplier and lot of enzyme.

Figure 3.
Description of the current 3SR amplification protocol using either the two–enzyme (AMV–RT) and T7 RNA polymerase) or three–enzyme formats (AMV RT, T7 RNA polymerase and *E. coli* RNase H) (Fahy *et al.*, 1991).

holds true must be empirically determined for each set of primers being used. The concentration of Mg^{2+}, deoxynucleotide and ribonucleotide triphosphates, and additives (sorbitol, dimethylsulfoxide) in the reaction buffer and the temperature of the reaction also can influence the rate of amplification (Fahy *et al.*, 1991) as is the case with PCR amplification (Varas *et al.*, 1991; Gelfand and White, 1990; Saiki, 1989; Innis and Gelfand, 1990).

Quantitation of initial target concentrations using the 3SR (or PCR) amplification reaction require that a direct relationship exists between input target and amplified product at assay time and target concentration being tested. *Apriori*, the target concentrations for clinical samples are not easily estimated,

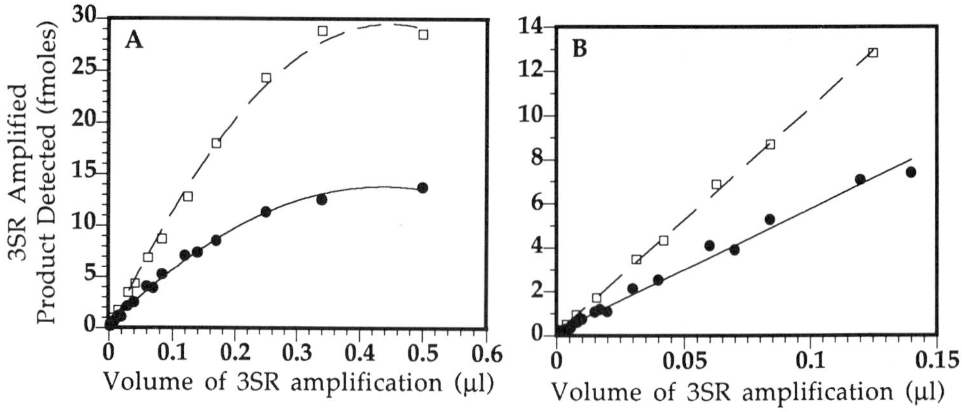

Figure 4.
Panel A: Influence of probe concentration on BBSH. A sample of 3SR product obtained by amplifying 1 attomole of HIV–1 RNA with the *env*–specific primers 88–211 (5′–AATTTAATACGACTCACTATAGGGATCTATTGTGCCCCGGCTG-GTTTTGCGATTCTA–3′) and 88–347 (5′–AATTTAATACGACTCACTATAG-GGATGTACTATTATGGTTTTAGCATTGTCTGTGA–3′) was diluted to give a range of concentrations corresponding to 0.5 μl or less of the 3SR reaction. Bead–based sandwich hybridizations (BBSH) were conducted using either 20 (solid circles) or 50 (open squares) femtomoles of 87–81 probe (5′–AATTAGGCCAGT-AGTATCAACTCAACT–3′) and Oligobeads™ 86–273 (5′–TGTACTATTATGG-TTTTAGCATTGTCTGTG–3′). Trisacryl Oligobead™ supports containing capture oligonucleotides used in BBSH are prepared as previously described (Guatelli *et al.*, 1990; Fahy *et al.*, 1991). Oligonucleotides are labeled at their 5′ ends using T4 polynucleotide kinase (Sambrook *et al.*, 1989). The 3SR amplification product is diluted in 10 mM Tris, pH 8.1, 1 mM EDTA, and added in a volume of 20 μl to 25 mg of Trisacryl Oligobeads™ 86–273 in a 2 ml microcolumn (Isolab). A 30 μl volume of prewarmed (42°C) hybridization solution (10X standard saline phosphate, EDTA, 20% (w/v) dextran sulfate and 0.2% SDS) is added to the column, followed by 10 μl (20 or 50 femtomoles of [^{32}P]–labeled detection oligonucleotide 87–81. The beads are then incubated at 42°C for 2 hours with occasional vortexing or continuous mechanical shaking and washed six times with 1 ml of 2X standard saline citrate at 42°C. The radioactivity bound to the beads or found in the combined washes is measured by Cerenkov counting. The amount of target detected is determsined by calculating the percentage of total counts bound to the beads and multiplying this quantity by the femtomole amount of labeled probe used in the assay.
Panel B: Expanded plot of the linear portion of Figure 4A.

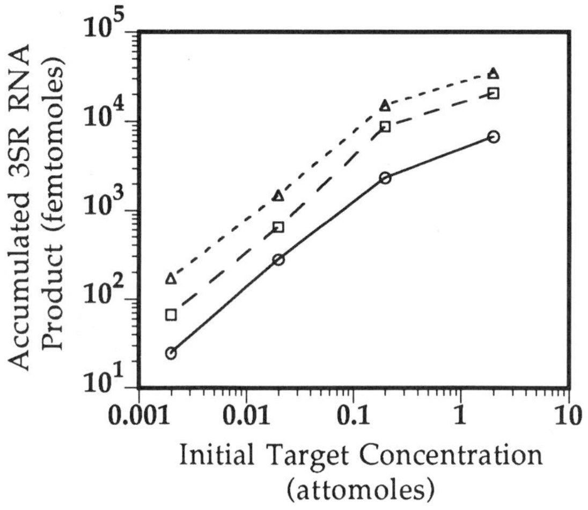

Figure 5.
The relationship between initial target concentration and 3SR product accumulation at various times in the amplification reaction of a 186 base region of the HIV–1 *env* gene. Target concentrations of 0.002, 0.02, 0.2 and 2 attomoles were amplified using the protocol outlined in Figure 3 and samples were assayed for accumulated 3SR RNA product by BBSH (as described in the legend to Figure 4A) at 20 (open circles), 25 (open squares) and 30 minutes (open triangles).

and thus for quantitation it is frequently necessary to titrate the target sequences in a clinical sample over a range which will include concentrations that allow for exponential increases during 3SR amplification of the target. Control sequences must be amplified in conjunction with the target sequence to allow distinction between failed reactions and absence of target sequence.

Control sequences used to test for amplification failure can also serve both to ensure the 3SR reaction is exponential and to calibrate the reaction. Several different types of standard sequences have been used as controls for PCR amplification reactions. The use of a target sequences which are of known copy number in the genome of cell containing samples, e.g., β–globin, has been one method used for standardization of PCR reactions (Davis *et al.*, 1990). Such DNA standards are not practical for the 3SR reaction if the reaction is used to amplify RNA targets. Use of a constitutively expressed cellular RNA, e.g., β_2–microglobin, is possible. However, such expression might vary either in different cell types or in infected cells, making this appraisal somewhat conditional. Additionally, the use of constitutively expressed RNA is not a method which could be used for samples such as serum or plasma (Morello *et al.*, 1985; Noonan *et al.*, 1990). For quantitation of PCR amplification reactions, several approaches have been employed including the use of predetermined amounts of an RNA or DNA standard (i.e., having a previously defined concentration) which are added as controls to the amplification reaction (Becker–André and Hahlbrock, 1989; Wang *et al.*, 1989; Gilliland *et al.*, 1990a; Gilliland *et al.*, 1990b; Ozawa *et al.*, 1990; Wang and Mark, 1990). Calibration curves which are generated in separate amplification reactions and used for the quantitation of PCR amplification assays also have been employed by several groups (Chelly *et al.*, 1988; Gaudette and Crain, 1991; Robinson and Simon, 1991). However, the PCR aided transcript titration assay (PATTY) described by Becker–André and Hahlbrock which uses an internally generated calibration curve may be the most promising approach for calibration (Becker–André and Hahlbrock, 1989).

In the PATTY method, amplification reactions containing the same size aliquots of target RNA are "spiked" with different predetermined amounts of a modified RNA standard. This control RNA contains the same sequences for PCR primer hybridization as the target RNA but has an inserted sequence in the middle of the amplified region such that the PCR product of the control RNA contains a restriction enzyme cleavage site which is not present in the target sequence or its PCR product. PCR amplifications are performed using a reverse transcriptase step at the beginning of the reaction. The DNA products from the target and RNA standard can be differentiated by cleavage of the standard by a restriction enzyme at the modified site. The target RNA is considered to be equal in concentration to the standard in the reaction when the amounts of standard and target PCR products are equivalent. Although the detection method would have to be altered, the PATTY approach to quantitation and amplification control could be applied directly to the 3SR reaction.

It is important to note that due to the sequence heterogeneity of HIV–1 isolates, a second type of control would seem to be necessary for amplification–based assays. Simultaneous coamplification of at least two regions of the HIV–1 genome should be performed to eliminate the possibility that amplification failure is due to lack of homology of one of the primers. The fact that the 3SR reaction is perform at 42°C makes this a less likely possibility than in the PCR reaction where amplification is performed at elevated temperatures. However, a significant lack of complementarity at the 3' end of a primer could slow or prevent 3SR cDNA synthesis leading to changes in the reaction kinetics or total amplification failure. Such 3SR coamplification reactions have been demonstrated involving the two regions of the HIV–1 *pol* gene which mutate when the virus becomes resistant to AZT. Currently, these 3SR coamplification reactions have produced amplification levels equivalent to the amplification levels obtained when the regions are amplified separately (E. Fahy, unpublished results).

3. Hybridization and detection

There are many approaches which could be utilized in the hybridization and detection steps of assays employing target amplification, and these are reviewed elsewhere (Thompson and Gillespie, 1990). Recently, two approaches to non–isotopic detection of 3SR amplified products appear to be particularly attractive. The first strategy is based on direct labelling of an oligonucleotide probe with a chelate moiety possessing a high affinity for rare earth metal cations, such as europium (Eu^{3+}) and terbium (Tb^{3+}). Following the BBSH reaction, the rare earth metal chelate (REMC) is detected by time–resolved fluorescence. Advantages of this detection approach are the minimization of background fluorescence due to the time–resolved mode of detection of the signal, and the large Stokes shift of the REMC moiety (Hemmilä *et al.*, 1984; Oser *et al.*, 1988). The second approach utilizes the oligonucleotide–enzyme conjugate probes, thus taking advantage of the signal amplification provided by the reporter enzyme (J. Ishii and S. Ghosh, unpublished results). Use of dioxetane–based substrates allows exquisite sensitivities for detection of the chemiluminescent products (Schaap, 1988; J. Ishii and S. Ghosh, unpublished results). Both strategies have been used to detect 3SR amplification products and are at least equivalent if not somewhat better in sensitivity to the [^{32}P]–BBSH assay and show the same range of linearity (C. Bush, D. Kwoh and S. Ghosh, unpublished results).

III. FACTORS WHICH WOULD FACILITATE USE OF AMPLIFICATION ASSAYS IN CLINICAL SETTINGS

A major concern in the use of target amplification–based assays is the occurrence of false positive reactions due to contaminating nucleic acid originating from carry–over amplified product from previous amplification using the same primers (Kwok and Higuchi, 1989; Kitchin *et al.*, 1990). This problem is especially troublesome in laboratories which use the same amplification primers repeatedly as would occur in a clinical reference laboratory. Due to the levels of amplification (typically 10^6–10^9–fold), transfer of even minute amounts of an amplification product from one tube to another may lead to false–positive reactions in tubes otherwise lacking target sequences. This so called "carry–over" contamination problem has been addressed in several recent reports describing "sterilization" techniques for PCR reactions which render the PCR amplification products unsuitable for further amplification (Longo *et al.*, 1990; Cimino *et al.*, 1991; Isaacs *et al.*, 1991; Ou *et al.*, 1991).

One method simply uses UV light to inactivate the amplified DNA by producing pyrimidine adducts (particularly thymidine bases). (Ou *et al.*, 1991). Formation of such adducts prevents the DNA from serving as a primer or template for DNA synthesis. The effectiveness of this approach is dependent on several factors such as the distance of the UV source from the sample and time of irradiation. This method adds another level of protection against false positive amplification reactions when used in conjunction with other procedural precautions (Kwok and Higuchi, 1989; Kitchin *et al.*, 1990).

The second sterilization strategy involves a pre–amplification enzymatic procedure (Longo *et al.*, 1990). In this protocol thymidine (T) residues are replaced by uracil (U) residues during PCR amplification. The bacterial enzyme uracil–N–glycosylase (UNG) (Duncan, 1991), which removes the uracil base from the phosphate backbone of the U–containing PCR product DNA but does not attack the T–containing target DNA, is utilized at the beginning of subsequent amplification reactions to eliminate any carryover PCR products. The UNG enzyme is subsequently inactivated during the first thermal denaturation step (95–97°C) in the PCR reaction, and the DNA is cleaved at the apyrimidine sites at these high temperatures eliminating it as a template for Taq DNA polymerase. The one consideration in the use of this procedure is the possibility that the UNG enzyme is not completely inactivated which concomitantly could lead to lowered amplification levels (Persing, 1991). Also substitution of dUTP for TTP in the PCR protocol can result in lower amplification efficiencies which must be made up by additional cycles of amplification.

A third sterilization strategy is a post–amplification procedure using cross–linking of the PCR products by UV light after incorporation of isopsoralen (4' amino–ethyl–4,5'–dimethylisopsoralen) into the DNA during amplification (Cimino *et al.*, 1991; Isaacs *et al.*, 1991). UV–irradiation of the reaction immediately following amplification activates the isopsoralens to form cyclobutane adducts with the pyrimidine nucleotides in both target and PCR product DNA. Such cross–linked DNA cannot serve as a template in PCR reactions if it happens to contaminate a subsequent amplification assays. Potential drawbacks to this procedure are focused around the fact that isopsoralen can interfere with efficient amplification and/or hybridization of detection probes. Consequently, careful titrations of isopsoralen is required for efficient inactivation of the PCR template but not inhibition of amplification and detection in reactions (Gingeras and Kwoh, 1991; Persing, 1991). Although no contamination control strategy has been reported for 3SR amplification as yet, the isopsoralen strategy could be applied to inactivate 3SR products.

Another area in which the amplification–based assays need to progress before they are widely accepted in a clinical setting is in automation of the procedures. This not only requires automation of the amplification reaction itself, but also the hybridization/detection steps and possibly the sample preparation steps. It is generally agreed that advances in this area require that the detection system be non–isotopic. Although several non–isotopic detection procedures have been reported for PCR–based assays (Persing, 1991), the data on the variability and range of quantitative detection are limited. As mentioned above, the non–isotopic detection systems of time–resolved fluorescence and chemiluminescence are being used in conjunction with 3SR amplification, but have not been developed to the point of automation as yet.

IV. CONCLUSIONS

Development of *in vitro* target amplification technologies has opened many new avenues of research in molecular diagnostics. PCR amplification is the most widely used of the amplification techniques and has set the standards for the field of amplification. The 3SR amplification system, although still in the early stages of development, has several unique characteristics which point the way to future applications. Because of its target selectivity (RNA versus DNA), the 3SR technology has the potential to allow low levels of gene expression to be detected and possibly quantitated. These features will aid in the understanding not only of viral diseases but also will benefit areas such genetic diseases, development, and in cancer research. Because of the single–stranded RNA products of the 3SR reaction, additional developments of this reaction could couple the 3SR amplification reaction to an *in vitro* translation system, thus allowing direct functionality testing of expressed genes. Finally, the application of the 3SR reaction to the detection of drug–resistant HIV–1 isolates has demonstrated the utility of the 3SR amplification/BBSH assay to detect and characterize genetic mutations. Identification of the locus of important genetic mutations remains the most problematic step in establishing such an assay. The combination of using the 3SR technology and the RNase mismatch scanning strategy (Myers *et al.*, 1985; Winter *et al.*, 1985; López–Galíndez *et al.*, 1991) offers an interesting solution to this problem.

ACKNOWLEDGMENTS

We thank S. Ghosh, J. Ishii, E. Fahy, and C. Bush for sharing unpublished results, S. Ghosh and J. Kwoh for critical review of the manuscript, and C. Vavrunek for preparation of the manuscript.

REFERENCES

Abbott, J. M., Poiesz, B. J., Byrne, B. C., Kwok, S., Sninsky, J. J. and Ehrlich, G. D. (1988) *J. Infect. Dis.* **158**, 1158–1169.

Barany, F. (1991a) *Proc. Natl. Acad. Sci. U.S.A.* **88**, 189–193.

Barany, F. (1991b) *PCR Method .Applic.* **1**, 5–16.

Barringer, K. J., Orgel, L., Wahl, G. and Gingeras, T. R. (1990) *Gene* **89**, 117–122.

Becker–André, M. and Hahlbrock, K. (1989) *Nucl. Acids Res.* **17**, 9437–9446.

Bhagavati, S., Ehrlich, G., Kula, R. W., Kwok, S., Sninsky, J., Udani, V. and Poeisz, B. J. (1988) *N. Engl. J. Med.* **318**, 1141–1147.

Boucher, C. A. B., Tersmette, M., Lange, J. M. A., Kellam, P., deGoede, R. E. Y., Mulder, J. W., Darby, G., Goudsmit, J. and Larder, B. A. (1990) *Lancet* **336**, 585–590.

Buchbinder, A., Josephs, S. F., Ablashi, D., Salahuddin, S. Z., Klotman, M. E., Manak, M., Krueger, G. R., Wong–Staal, F. and Gallo, R. C. (1988) *J. Virol. Meth.* **21**, 191–197.

Chelly, J., Kaplan, J.–C., Maire, P., Gautron, S. and Kahn, A. (1988) *Nature* **333**, 858–860.

Chou, S. and Merigan, T. C. (1982) *N. Engl. J. Med.* **308**, 921–925.

Cimino, G. D., Metchette, K. C., Hearst, J. E. and Isaacs, S. T. (1991) *Nucl. Acids Res.* **19**, 99–107.

Davis, G. R., Blumeyer, K., DiMichele, L. J., Whitfield, K. M., Chappelle, H., Riggs, N., Ghosh, S. S., Kao, P., Fahy, E., Kwoh, D. Y., Guatelli, J. C., Spector, S. A., Richman, D. D. and Gingeras, T. R. (1990) *J. Infect. Dis.* **162**, 13–20.

de la Salle, C., Bass, M. J., Laustriat, D., Guy, B., Bordes, E., Wiesel, M. L., Grunebaum, L., Lutz, P., Partisani, M., Lang, J. M. *et al.*(1991) *Ann. Hematol.* **62**, 165–168.

Demmler, G. J., Buffone, G. J., Schimbor, C. M. and May, R. A. (1988) *J. Infect. Dis.* **158**, 1177–1184.

Duggan, D. B., Ehrlich, G. D., Davey, F. P., Kwok, S., Sninsky, J., Goldberg, J., Baltrucki, L. and Poiesz, B. J. (1988) *Blood* **71**, 1027–1032.

Duncan, B. K., (1991) In *The Enzymes* (P. Boyer, ed.), pp. 565–586, Academic Press, Inc., New York.

Erlich, H. A., ed. (1989). In *PCR Technology: Principles and Applications for DNA Amplification*, Stockton Press, New York.

Erlich, H. A., Gelfand, D. H. and Saiki, R. K. (1988) *Nature* **331**, 461–472.

Fahy, E., Kwoh, D. Y. and Gingeras, T. R. (1991) *PCR Method. Applic.* **1**, 25–33.

Farzadegan, H., Polis, M. A., Wolinsky, S. M., Rinaldo, C. R. J., Sninsky, J. J., Kwok, S., Griffith, R. L., Kaslow, R. A., Phair, J. P., Polk, E. B. and Saah, A. J. (1988) *Ann. Intern. Med.* **108**, 785–790.

Gama, R. E., Hughes, R. J., Bruce, C. B. and Stanway, G. (1988) *Nucl. Acids Res.* **16**, 9346.

Gaudette, M. F. and Crain, W. R. (1991) *Nucl. Acids Res.* **19**, 1879–1884.

Gelfand, D. H. and White, T. J., (1990) In *PCR Protocols: A Guide to Methods and Applications* (M. Innis, D. Gelfand, J. Sninsky and T. White, eds.), pp. 129–141, Academic Press, New York.

Gibbs, R. A., Chamberlain, J. S. and Caskey, C. T., (1989) In *PCR Technology: Principles and Applications for DNA Amplification* (H. Erlich, ed.), pp. 171–191, Stockton Press, New York.

Gilliland, G., Perrin, S. and Bunn, H. F., (1990a) In *PCR Protocols: A Guide to Methods and Applications* (M. Innis, D. Gelfand, J. Sninsky and T. White, eds.), pp. 60–69, Academic Press, Inc., San Diego, CA.

Gilliland, G., Perrin, S., Blanchard, K. and Bunn, H. F. (1990b) *Proc. Natl. Acad. Sci. U.S.A.* **87**, 2725–2729.

Gingeras, T. R. and Kwoh, D. Y. (1991) In *Diagnostic Nucleic Acid Probe Technology* (C. Dangler and B. Osburn eds.) Hanser Publishers, New York.

Gingeras, T. R., Prodanovich, P., Latimer, T., Guatelli, J. C., Richman, D. D. and Barringer, K. J. (1991) *J. Infect. Dis.* **164**,

Gingeras, T. R., Richman, D. D., Kwoh, D. Y. and Guatelli, J. C. (1990a) *Veterinary Microbiology* **24**, 235–251.

Gingeras, T. R., Whitfield, K. M. and Kwoh, D. Y. (1990b) *Annales du Biologie Clinique* **48**, 498–501.

Guatelli, J. C., Whitfield, K. M., Kwoh, D. Y., Barringer, K. J., Richman, D. D. and Gingeras, T. R. (1990) *Proc. Natl. Acad. Sci. U.S.A.* **87**, 1874–1878.

Gyllensten, U. and Erlich, H. A. (1988) *Proc. Natl. Acad. Sci. U.S.A.* **85**, 7652–7656.

Hart, C., Schochetman, G., Spira, T., Lifson, A., Moore, J., Galphin, J., Sninsky, J. and Ou, C. Y. (1988) *Lancet* **2**, 596–599.

Hemmilä, I., Dakuba, S., Mukkala, V.–M., Sitari, H. A. and Lövgren, T. (1984) *Anal. Biochem.* **137**, 335–343.

Hewlett, I., Y., L., Epstein, J., Hawthorne, C. A., Ruta, M. and Allain, J. P. (1990) *J. AIDS* **3**, 714–720.

Imagawa, D. T., Lee, M. H., Wolinsky, S. M., Sano, K., Morales, F., Kwok, S., Sninsky, J.J., Nishanian, P.G., Giorgi, J., Fahey, J.L., Dudley, J., Visscher, B.R. and Detels, R. (1989) *N. Engl. J. Med.* **320**, 1458–1462.

Innis, M. A. and Gelfand, D. H., (1990). In *PCR Protocols: A Guide to Methods and Applications*, (M. Innis, D. Gelfand, J. Sninsky and T. White, eds.). Academic Press, Inc., San Diego.

Isaacs, S. T., Tessman, J. W., Mitchette, K. C., Hearst, J. E. and Cimino, G. D. (1991) *Nucl. Acids Res.* **19**, 109–116.

Jackson, J. B., Kwok, S. Y., Sninsky, J. J., Hopsicker, J. S., Sannerud, K. J., Rhame, F. S., Henry, K., Simpson, M. and Balfour, H. H. J. (1990) *J. Clin. Microbiol.* **28**, 16–19.

Japour, A. J., Chatis, P. A., Eigenrauch, H. A. and Crumpacker, C. S. (1991) *Proc. Natl. Acad. Sci. U.S.A.* **88**, 3092–3096.

Kitchin, P. A., Szotyeri, Z., Fromhole, C. and Almond, N. (1990) *Nature* **344**, 201.

Kwoh, D. Y., Davis, G. R., Whitfield, K. M., Chappelle, H. L., DiMichele, L. J. and Gingeras, T. R. (1989) *Proc. Natl. Acad. Sci. U.S.A.* **86**, 1173–1177.

Kwoh, D. Y. and Kwoh, T. J. (1990) *Am. Biotech. Lab.* **8**, 16–25.

Kwok, S. and Higuchi, R. (1989) *Nature* **339**, 237–238.

Kwok, S., Kellogg, D., Ehrlich, G., Poiesz, B., Bhagavati, S. and Sninsky, J. J. (1988) *J. Infect. Dis.* **158**, 1193–1197.

Landry, M. L. (1990) *Clin. Biochem.* **23**, 267–277.

Larder, B. A., Chesebro, B. and Richman, D. D. (1990) *Antimicrob. Agents Chemother.* **34**, 436–441.

Larder, B. A., Darby, G. and Richman, D. D. (1989) *Science* **243**, 1731–1734.

Larder, B. A. and Kemp, S. D. (1989) *Science* **246**,

Larzul, D., Guigue, F., Sninsky, J. J., Mack, D. H., Bréchot, C. and Guesdon, J. L. (1988) *J. Virol. Meth.* **20**, 227–237.

Laure, F., Courgnaud, V., Rouzioux, C., Blanche, S., Veber, F., Burgard, M., Jacomet, C., Griscelli, C. and Bréchot, C. (1988) *Lancet* **2**, 538–541.

Lifson, A. R., Stanley, M., Pane, J., O'Malley, P. M., Wiber, J. C., Stanley, A., Jeffrey, B., Rutherford, G. W. and Sohmer, P. R. (1990) *J. Infect. Dis.* **161**, 436–439.

Lizardi, P. M., Guerra, C. E., Lomeli, H., Tussie–Luna, I. and Kramer, F. R. (1988) *BioTechnology* **6**, 1197–1201.

Longo, M. C., Berninger, M. S. and Hartley, J. L. (1990) *Gene* **93**, 125–128.

López–Galíndez, C., Rojas, J. M., Nájera, R., Richman, D. D. and Perucho, M. (1991) *Proc. Natl. Acad. Sci. U.S.A.* **88**, 4280–4284.

McCabe, P. C., (1990) In *PCR Protocols: A Guide to Methods and Applications* (M. Innis, D. Gelfand, J. Sninsky and T. White, eds.), pp. 76–83, Academic Press, Inc., San Diego, CA.

Morello, D., Duprey, P., Israel, A. and Babinet, C. (1985) *Immunogenetics* **22**, 441–452.

Mullis, K. B. (1990) *Scientific American* **262**, 56–61, 64–65.

Mullis, K. B. and Faloona, F. A. (1987) *Methods Enzymol.* **155**, 335–350.

Murakawa, G. J., Zaia, J. A., Apallone, P. A., Stephens, D. A., Kaplan, B. E., Wallace, R. B. and Rossi, J. J. (1988) *DNA* **7**, 287–295.

Myers, R. M., Laren, Z. and Maniatis, T. (1985) *Science* **230**, 1242–1246.

Noonan, K. E., Beck, C., Holzmayer, T. A., Chin, J. E., Wunder, J. S. andrulis, I. L., Gazdar, A. F., William, C. L., Von Hoff, D. D. and Robinson, I. B. (1990) *Proc. Natl. Acad. Sci. U.S.A.* **87**, 7160–7164.

Oser, A., Roth, W. K. and Valet, G. (1988) *Nucl. Acids Res.* **16**, 1181–1196.

Ou, C.–Y., Kwok, S., Mitchell, S. W., Sninsky, J. J., Krebs, J. W., Feorino, P., Warfield, D. and Schochetman, G. (1988) *Science* **239**, 295–297.

Ou, C.–Y., Moore, J. L. and Schochetman, G. (1991) *BioTechniques* **10**, 442–445.

Ozawa, T., Tanaka, M., Ikebe, S.–I., Ohno, K., Kondo, T. and Mizuno, Y. (1990) *Biochem. Biophys. Res. Comm.* **172**, 483–489.

Paterlini, P., Gerken, G., Nakajima, E., Terre, S., D'Errico, A., Grigioni, W., Nalpas, B., Franco, D., Wands, J., Kew, M., Pisi, E., Tiollais, P. and Bréchot, C. (1990a) *N. Engl. J. Med.* **323**, 80–85.

Paterlini, P., Lallemant–Le Coeur, S., Lallemant, M., M'Pele, P., Dazza, M. C., Terre, S., Moncany, M., Jourdain, G., Courgnaud, V. and N'Zingoula, S. (1990b) *J. Med. Virol.* **30**, 53–57.

Persing, D. H. (1991) *J. Clin. Microbiol.* **29**, 1281–1285.

Persing, D. H. and Landry, M. L. (1989) *Yale J. Biol. Med.* **62**, 159–171.

Pezzella, M., Rossi, P., Lombardi, V., Gemelli, V., Mariani Constantini, R., Mirolo, M., Fundaro, C., Moschese, V. and Wigzell, H. (1989) *BMJ* **298**, 713–716.

Rayfield, M., DeCork, K., Heyward, W., Goldstein, L., Krebs, J., Kwok, S., Lee, S., McCormick, J., Moreau, J. M., Odebouri, K., Schochetman, G., Sninsky, J. and Ou, C. Y. (1988) *J. Infect. Dis.* **158**, 1170–1176.

Richman, D., Rosenthal, A. S., Skoog, M., Eckner, R. J., Chou, T.-C., Sabo, J. P. and Merluzzi, V. J. (1991b) *Antimicrob. Agents Chemotherap.* **35**, 305–308.

Richman, D. D. (1990) *Rev. Infectious Diseases* **12**, S507–S512.

Richman, D. D., Grimes, J. M. and Lagakos, S. W. (1990) *J. AIDS* **3**, 743–746.

Richman, D. D., McCutchan, J. A. and Spector, S. A. (1987) *J. Infect. Dis.* **156**, 823–827.

Robinson, M. O. and Simon, M. I. (1991) *Nucl. Acids Res.* **19**, 1557–1562.

Rogers, M. F., Ou, C.-Y., Rayfield, M., Thomas, P. A., Schoenbaum, E. E., Abrahms, E., Krasinski, K., Selwyn, P. A., Moore, J., Kaul, A., Grimm, K. T., Bamji, M. and Schochetman, G. (1989) *N. Engl. J. Med.* **320**, 1649–1654.

Saiki, R. K., (1989). *PCR Technology: Principles and Applications for DNA Amplification*, Stockton Press, New York.

Saiki, R. K., Gelfand, D. H., Stoffel, S., Scharf, S. J., Higuchi, R., Horn, G. T., Mullis, K. B. and Erlich, H. A. (1988) *Science* **230**, 1350–1354.

Saiki, R. K., Scharf, S., Faloona, F. A., Mullis, K. B., Horn, G., Erlich, H. and Arnheim, W. (1985) *Science* **230**, 1300–1354.

Sambrook, J., Fritsch, E. F. and Maniatis, T., (1989). *Molecular Cloning: A Laboratory Manual*, Cold Spring Harbor Laboratory Press, Cold Spring Harbor, NY.

Sheppard, H. W., Ascher, M. S., Busch, M. B., Sohmer, P. R., Stanley, M., Luce, M. C., Chimera, J. A., Madej, R., Rodgers, G. C., Lynch, C. and The New York City Collaborative Study of maternal HIV Transmission and Montefiore Medical Center HIV Perinatal Transmission Study Group (1991) *J. AIDS* **4**, 277–283.

Shibata, D., Fu, Y. S., Gupta, J. W., Shah, K. V., Arnheim, N. and Martin, W. J. (1988a) *Lab. Invest.* **59**, 555–559.

Shibata, D., Martin, W. J., Appleman, M. D., Causey, D. M., Leedom, J. M. and Arnheim, N. (1988b) *J. Infect. Dis.* **158**, 1185–1192.

Shibata, D. K., Arnheim, N. and Martin, W. J. (1988c) *J. Exp. Med.* **167**, 225–230.

Thompson, J. D. and Gillespie, W. R. (1990) *Clin. Biochem.* **23**, 261–265.

Varas, F., Leandro, M., Ballester, S. and Nájera, R. (1991) *BioTechniques* **11**, 384–391.

Wang, A. M., Doyle, M. V. and Mark, D. F. (1989) *Proc. Natl. Acad. Sci. U.S.A.* **86**, 9717–9721.

Wang, A. M. and Mark, D. F., (1990) In *PCR Protocols: A Guide to Methods and Applications* (M. Innis, D. Gelfand, J. Sninsky and T. White, eds.), pp. 70–75, Academic Press, Inc., San Diego, California, USA.

Wedell, A., Datta, S. andersson, B. and Luthman, H. (1990) *Technique* **2**, 23–26.

Williams, P., Simmonds, P., Yap, P. L., Balfe, P., Bishop, J., Brettle, R., Hague, R., Hargreaves, D., Inglis, J. and Brown, A. L. (1990) *AIDS* **4**, 393–398.

Winter, E., Yamamoto, F., Almoguera, C. and Perucho, M. (1985) *Proc. Natl. Acad. Sci. U.S.A.* **82**, 7575–7579.

Wolinsky, S. M., Rinaldo, C. R., Kwok, S., Sninsky, J. J., Gupta, P., Imagawa, D., Farzadegan, H., Jacobson, L. P., Grovit, K. S. and Lee, M. H. (1989) *Ann. Intern. Med.* **111**, 961–972.

Wu, D. Y. and Wallace, R. B. (1989) *Genomics* **4**, 560–569.

Zeldis, J. B., Lee, J. H., Mamish, D., Finegold, D. J., Sircar, R., Ling, Q., Knudsen, P. J., Kuramoto, I. K. and Mimms, L. T. (1989) *J. Clin. Invest.* **84**, 1503–1508.

Zignego, A. L., Deny, P., Feray, C., Ponzetto, A., Gentilini, P., Tiollais, P. and Bréchot, C. (1990) *Mol. Cell. Probes* **4**, 43–51.

4

The Molecular Genetics of Hepatitis C Virus (HCV) and its Use in the Diagnosis of HCV Infections

Edouard Kurstak, Christine Kurstak and A. Hossain

I. INTRODUCTION

The genome of one of the causative agents of non–A, non–B (NANB) hepatitis was discovered by Choo *et al.*, (1989) using re–combinant complementary DNA (cDNA) technology. This agent, now known as hepatitis C virus (HCV), is known to be the predominant cause of post–transfusion NANB hepatitis worldwide (Alter *et al.*, 1989; Esteban *et al.*, 1989; Kuo *et al.*, 1989; Van der Poel *et al.*, 1989; Choo *et al.*, 1990; Kurstak, 1993). Anti–HCV has been associated with most cases of hepatocellular carcinoma (HCC) in Italy (Colombo *et al.*, 1989), Spain (Bruix *et al.*, 1989) and Japan (Nishioka *et al.*, 1991). HCV thus appears to be a major etiological agent of NANB hepatitis, chronic liver disease and HCC around the globe. This review focuses on the molecular biology of HCV and its utilization in the diagnosis of hepatitis C virus disease.

II. HEPATITIS C VIRUS (HCV) GENOME: STRUCTURE AND FUNCTION

HCV contains a positive–strand RNA genome comprising about 9400 nucleotides (nt; Choo *et al.*, 1989; Takamizawa *et al.*, 1991). This sequence contains a single, large translation open–reading frame (ORF) that spans almost the entire genome and could encode a large viral polypeptide of either 3,011 (Choo *et al.*, 1991) or 3,010 (Takamizawa *et al.*, 1991) amino acids beginning with the first methionine codon of the ORF. The 5'–terminal region represents the

---

Edouard Kurstak, Christine Kurstak, and A. Hossain • Department of Microbiology and Immunology, Faculty of Medicine, University of Montreal, Montreal, Quebec, Canada

Applied Virology Research, Vol. 3, Edited by
E. Kurstak *et al.*, Plenum Press, New York, 1994

most highly conserved sequence among different viral isolates (Choo *et al.*, 1991; Han *et al.*, 1991; Okamoto *et al.*, 1991), thus suggesting that it may have an important regulatory role during viral replication, very likely at the level of translation. The longest 5'–terminal region thus far identified consists of 341 nucleotides, which may represent most of the 5' genome terminus and perhaps a small hairpin secondary structure at the 5' terminus (Han *et al.*, 1991).

Downstream from the large ORF, a small presumably 3' untranslated sequence has been identified and it is known to contain between 27 and 55 nt, depending on the source of the virus (Choo *et al.*, 1991; Takamizawa *et al.* , 1991). HCV RNA binds to oligo (dT) cellulose and thus contains a poly (A) tract, in all likelihood at the 3' terminus (Choo *et al.*, 1989).

Very little overall homology exists between the sequence of both the HCV genome and its encoded polypeptide and other known viral sequences , evidence for HCV being a novel type of virus. Several lines of evidence indicate HCV as a being distantly related to the human flaviviruses (yellow fever virus, Dengue viruses and the Japanese encephalitis virus) and animal pestiviruses (bovine viral diarrhea virus) currently recognized as members of the same Flaviviridae family (Collett *et al.*, 1988). A computerized comparative study of the sequence of a HCV–encoded polypeptide has indicated that a small region of the HCV polypeptide shows significant sequence homology with the nucleoside triphosphate– binding helicases encoded by flaviviruses, plant potyviruses and pestiviruses, with the latter having the greatest sequence identity with HCV (Miller & Purcell, 1990; Choo *et al.*, 1991). Furthermore, other regions of amino acid sequence have also shown identity between HCV and the serine proteases and RNA–dependent RNA polymerase encoded by flaviviruses, pestiviruses and plant viruses (Miller & Purcell, 1990; Choo *et al.*, 1991; Takamizawa *et al.*, 1991). However, these regions of polypeptide sequence similarity appear to be colinear only in cases of HCV, the flaviviruses and the pestiviruses, indicative of a similarity in genetic and polypeptide organization (Choo *et al.*, 1991). The observation that very little primary sequence identity exists between these viruses in terms of their hydropathicity profiles (Choo *et al.*, 1991; Takamizawa *et al.*, 1991), lends support to the idea of three virus types with an apparent similarity in structure and organization of encoded polypeptides. In addition, each of these viral polypeptides is known to contain a basic nucleocapsid protein domain at the N'–terminus followed by glycoprotein domains (Miller & Purcell, 1990; Choo *et al.*, 1991; Takamizawa *et al.*, 1991).

At the nucleotide sequence level, very little homology has been shown to exist with any known virus except at the 5'–terminal region upstream from the sequence encoding the large viral glycoprotein. A high extent of homology is noticeable between this region and the analogous region in the pestiviruses (Choo *et al.*, 1991, Han *et al.*, 1991). Additionally, this nucleotide sequence similarity with the pestiviruses suggests that HCV may be closer in evolution to the pestiviruses than the flaviviruses. Furthermore, the extensive homology observed with the pestiviruses at the level of the primary polypeptide sequence appears to be consistent with this hypothesis. However, the hydropathicity of the HCV polypeptide seems to better match that of the flaviviral polyprotein than that of the pestiviruses (Choo *et al.*, 1991). Comparative sequence analyses suggests that the pestiviruses and flaviviruses are the nearest relatives of HCV, data which is also consistent with earlier filtration and solvent inactivation data indicating that HCV is a small, enveloped virus (He *et al.*, 1987). This prompts the classification of these three virus types as three separate genus in the Flaviviridae family, and finds credence in the recent assignment of the pestiviruses to this family (Collett *et al.*, 1988).

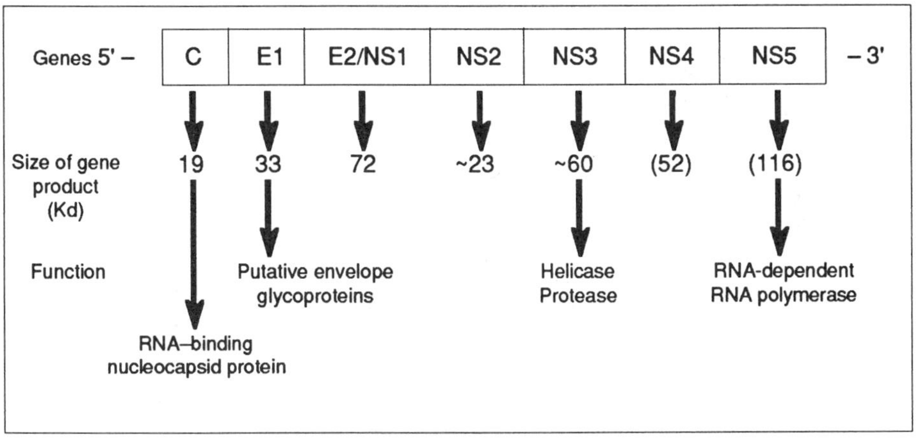

C : Nucleocapsid protein gene
E1 : Envelope gene
E2/NS1 : Envelope gene or non–structural gene (NS1)
NS2 : Non–structural gene 2
NS3 : Non–structural gene 3
NS4 : Non–structural gene 4
NS5 : Non–structural gene 5

Figure 1. Genetic Organization of the Hepatitis C Virus (HVC).

A schematic illustration of the genetic organization of HCV is shown in Figure 1. Similar to pestiviruses and flaviviruses, HCV appears to encode a large polyprotein precursor from which individual viral proteins (structural as well as non–structural) are processed co–translationally and/or post–translationally through the combined action of host–encoded and viral–encoded proteases. Virion structural proteins are processed from approximately the C–terminal quarter of the polyprotein with a variety of non–structural (NS) proteins being processed from the remainder of the polyprotein. Viral proteins and particles as yet remain unidentified in HCV–infected individuals. It has, however, only been possible to identify viral proteins by transcribing RNA from cloned HCV cDNA *in vitro* and translating the RNA in *in vitro* translation systems. In view of the observed genetic relationship between HCV and the pestiviruses and flaviviruses, gp33 may correspond to a matrix/envelope glycoprotein in the virion, whereas gp72 may correspond to a viral envelope glycoprotein equivalent to gp53/gp55 of the pestiviruses (Collett *et al.*, 1989) or represent the first NS protein equivalent to the NS1 glycoprotein of the flaviviruses (Rice *et al.*, 1986).

The function of NS2 remains to be elucidated but the protein is very hydrophobic (Choo *et al.*, 1991). Recent expression analysis of HCV cDNA transfected into Chinese hamster ovary cells indicates that when synthesized from a precursor, HCV E2/NS1 is not normally secreted into the cell media but like E1, is retained within the endoplasmic reticulum. This behaviour of E2 resembles that of the pestivirus envelope counterpart (gp53–55) rather than the flaviviral NS1 molecule, which is a secreted protein (Rice *et al.*, 1986; Stark *et al.*, 1990; Collett *et al.*, 1991). The NS3 protein contains a nucleoid triphosphate–binding helicase enzyme most likely involved in processing the NS proteins from the polyprotein precursor (Choo *et al.*, 1991). The products of HCV NS4 and NS5 genes have not yet been characterized. In the flaviviruses, two proteins are encoded by each of the NS2 and NS4 genes (NS2 a + b and NS4 a + b) but the situation in HCV remains unknown (Chambers *et al.*, 1990). Although the NS5 gene product may be multi–functional, this protein contains the RNA–dependent RNA polymerase that is involved in the replication of the RNA genome. Like the pestiviruses and flaviviruses, HCV does not appear to produce DNA replication intermediates, and integrated forms of the viral genome in the host genome have not been detected (Choo *et al.*, 1989). However, use of the polymerase chain reaction (PCR) method has provided evidence of subgenomic RNA species (Han *et al.*, 1991) in contrast to the absence of such in the replication of the flaviviruses and pestiviruses (Rice *et al.*, 1986; Collett *et al.*, 1989).

III. DIVERSE NATURE OF HEPATITIS C VIRUSES

An apparent diversity in the nucleotide sequence of different HCV isolates is now obvious. Comparison of the sequence of HCV–J1 , a Japanese isolate, with the sequence of the original isolate derived in the United States. (HCV–1; Choo *et al.*, 1989) revealed that HCV–J1 differed in nt polypeptide sequences in the NS3 and NS4 regions (Kubo *et al.*, 1989). This conclusion was later extended to the NS5 (Takeuchi *et al.* , 1990a) and envelope regions (Takeuchi *et al.*, 1990b) and is suggestive of the existence of multiple types of HCV as in the case of flaviviruses and pestiviruses. Currently, the complete coding sequence in addition to 5'–terminal and 3'–terminal sequences are available for HCV–1 isolated in the United States (Choo *et al.*, 1991) and two independent isolates from Japan (Kato *et al.*, 1990). The Japanese isolates have been noted to be very similar in terms of both total nt and total polypeptide sequence homology but less so in total sequence similarity with HCV–1 .

Comparative sequence analysis of the complete and partial HCV sequences known to date (actual nucleotide and AA sequence) indicates four basic genotype groups HCV–I, HCV–II, HCV–III, HCV–IV, which can be further segregated into two core antigen subtypes (Machida *et al.*, 1992; Okamoto *et al.*, 1992). HCV genotypes are known to differ in different countries (Takada *et al.*, 1992). A greater sequence similarity among proteins of the same virus group than between the different groups has been noted. Substantial amino acid sequence variation between groups I and II in the putative viral envelope proteins encoded by the E1 and E2/NS1 genes is also apparent. The region corresponding to the N–terminal 30 amino acids of E2/NS1 shows substantial variation among nearly all isolates, independent of the virus group (Hijikata *et al.*, 1991; Weiner *et al.*, 1991). Heterogeneity observed between different HCV isolates may be of importance in virus/host interactions, evolution of chronicity and vaccine development.

An important clinical aspect of HCV diversity is the probability of multiple infection with different HCV agents. In a pioneering investigation, it was observed that of 14 Japanese patients with chronic NANB hepatitis all had detectable levels of circulating, presumed group I, genome but in addition four had circulating group III genome. Group I, II, III and IV viruses have all been observed in Japan (Kato *et al.*, 1990; Okamoto *et al.*, 1991; Takamizawa *et al.*, 1991; Machida *et al.*, 1992). Multiple HCV genotypes are found in Europe (Simmonds *et al.*, 1990). Over a dozen different genotypes of HCV have been described thus far, and these can be divided into at least six major groups. The minor groups may differ from each other in genetic heterogeneity by upto 10%, whereas major gorups may differ by as much as 50% (Purcell, 1993). Certain regions of the genome appear to be more variable in sequence, and at least two hypervariable regions have been identified in the E2 gene (Purcell, 1993). It is likely that each separate isolate of HCV is genetically unique.

In a recent study of 51 isolates of HCV collected worldwide, Bukh *et al.* (1993) predicted, by sequence analysis of the putative E1 gene, at least 12 distinct genotypes of hepatitis C virus. The E1 gene sequence of 8 of these genotypes has not been reported in previous studies. They found that some genotypes of HCV are widely distributed worldwide, whereas others were identified only in defined geographical regions. For example, 4 genotypes were identified exclusively in Africa and comprised the majority of hepatitis C virus isolates on that continent. In all 51 isolates of HCV, the E1 gene was exactly 576 nucleotides in length with no in–frame stop codons. Several conserved domains important in maintaining biological function were identified in the predicted E1 protein: 8 invariant cysteine residues; 3 potential N–linked glycosylation sites; a domain of 9 amino acids (GHRMAWDMM); and an amino acid doublet near the putative cleavage site at the C terminus of the protein (Bukh *et al.*, 1993).

The evidence of heterogeneity of at least 12 genotypes of hepatitis C virus has important implications for HCV diagnosis, in virus/host interactions, evolution of chronicity of infection and vaccine development.

IV. MOLECULAR GENETICS AND DIAGNOSIS OF HCV INFECTIONS

Unlike the situation found with the flaviviruses and pestiviruses and other viral infections, information on the molecular genetics of HCV largely preceded knowledge of its virology, immunology and biology.

Tremendous progress has been made towards elucidating the structure and organization of the genome of HCV isolated from human carriers. The availability of sequence information as well as cDNA spanning the entire coding region have provided the impetus for virological studies of HCV, the development of valuable diagnostic tools and vaccines to prevent the majority of post–transfusion NANB hepatitis cases andHCV–associated malignant liver diseases.

A. Immunological Diagnosis of HCV Infections

The molecular characterization and cloning of HCV rapidly led to the production of large quantities of viral protein from recombinant organisms. On purification, the recombinant polypeptide could be effectively used to produce a sensitive capture assay for reactive antibody, indicating exposure of the individual to the hepatitis C virus. The first of such assays incorporated the C100–3 polypeptide expressed in yeast (Kuo *et al.*, 1989). In this recombinant assay, three overlapping clones were isolated by means of the cDNA in HCV clone 5–1–1 which was used as a hybridization probe to the original cDNA library. The clones have a common open reading frame (ORF) extending throughout that encodes part of a viral antigen associated with NANB hepatitis (Choo *et al.*, 1989). This continuous ORF was reconstructed from the clones and then expressed in yeast as a fusion polypeptide with human superoxide dismutase (SOD), which facilitates the efficient expression of foreign proteins in yeast and bacteria. A SOD/HCV polypeptide (C100–3) was synthesized in recombinant yeast. C100–3 was used to coat the wells of microtitre plates on which circulating HCV antibodies in blood samples could be captured and measured. Detection of bound antibody was achieved with a radioactive second antibody. The C100–3 is currently envisioned to correspond to nearly all of the NS4 protein of the HCV–1 isolate (Choo *et al.*, 1989, 1991; Kuo *et al.*, 1989). Furthermore, antibody to this specific protein has been noted to develop in most cases of post–transfusion NANB hepatitis (Alter *et al.*, 1989; Colombo *et al.*, 1989; Esteban *et al.*, 1989, Kuo *et al.*, 1989). In particular, the antibody appears to be associated with chronic infections (Alter *et al.*, 1989; Tanaka *et al.*, 1991) and is a good marker of the presence of infectious virus (Esteban *et al.*, 1990; Weiner *et al.*, 1990). It has proved to be an efficient assay for NANB hepatitis infections worldwide, and apparently precedes detection of infection by HCV groups I and II (Katayawa *et al.*, 1990; Takeuchi *et al.*, 1990; Takamizawa *et al.*, 1991). Failure to detect circulating antibodies to C–100 antigen has, however, been observed in some patients with post–transfusion NANB hepatitis (Skidmore, 1990) and the test has been found to give rise to false–positive results (McFarlane *et al.*, 1990). Perhaps, using additional recombinant–derived viral proteins to detect other antibodies, a more sensitive assay for anti–HCV could be produced as well as one that reduces the time for detection of seroconversion. The use of viral proteins well conserved between different HCV groups (NS3, NS4) might hopefully ensure a broadly reactive and sensitive diagnostic tool. In fact, new assays incorporating epitopes identified in recombinant antigens derived from the nucleocapsid (C) as well as the NS3 and NS4 domains (C22c, C33c, C200) have been developed. These assays have served to increase the sensitivity of detection of HCV antibodies as well as allowing the early diagnosis of HCV infection (Alberti, 1991; Van der Poel *et al.*, 1991).

B. Detection of HCV–RNA

The molecular characterization of the HCV genome has led to the development of immunoassays for the detection of antibodies to HCV. The first–generation enzyme immunoassay (EIA) basically used a single recombinant viral protein, designated C100–3, for the diagnosis of HCV infection (Kuo *et al.*, 1989).

Use of the C100–3 EIA resulted in a significant decrease in transfusion associated hepatitis infections (Donahue *et al.*, 1992). However, it now appears clear from a number of studies that the C100–3 EIA generated significant numbers of false positive and false negatives results (Follett *et al.*, 1991; Okamoto *et al.*, 1991, Sugitani *et al.*, 1992), the latter being specifically associated with blood donors without elevated levels of alanine aminotransferase (ALT). To ameliorate these problems, second and third–generation serological tests that include additional structural and non–structural protein or peptide epitopes are currently being evaluated in screening as well as confirmatory assays (Follett *et al.*, 1991; Boudart *et al.*, 1992; Okamoto *et al.*, 1992). These tests are similar to the first generation test in that each relies on detection of an antibody response to viral antigens. Okamoto *et al.*, (1992) developed immunoassays to detect antibodies against oligopeptides deduced from the putative core gene of hepatitis C virus, and compared their performance with that of a commercial immunoassay for antibodies against the product of non–structural regions of HCV (anti–C100–3). A 19–mer oligopeptide (CP10) and a 36–mer oligopeptide (C9) were chemically synthesized representing hydrophilic regions of the product of the HCV core gene. THe oligopeptides were used to capture corresponding antibodies anti–CP10 and anti–CP9, by enzyme–linked immunosorbent assay, in sera from patients with acute or chronic NANB hepatitis and in blood donations. In patients with chronic NANB hepatitis the frequency of detection of anti–CP9, anti–CP10 or both, was noted to be higher (91%) in comparison to anti–C100–3 (74%). Among sera positive by at least one test, 14 were found to contain hepatitis C virus RNA; 7 of which were negative for anti–C100–3 but positive for anti–CP10 and/or anti–CP9 in high titers. These results seem to suggest that antibodies against antigenic determinants of the hepatitis C virus core could possibly complement anti–C100–3 in the reliable diagnosis of NANB hepatitis and thereby further decrease the incidence of post–transfusion NANB hepatitis.

Because the period between infection and host–immune response to different HCV antigens can vary among individuals, it would perhaps be desirable to test directly for viral antigens, nucleic acids or even both. However, because of the low titers of circulating virus that are common in HCV infection (Ulrich *et al.*, 1990) reliable detection of viral antigens remains as yet to be documented. Consequently, extensive efforts have been directed towards the detection of HCV RNA in clinical samples.

The application of polymerase chain reaction (PCR) techniques to amplify reverse–transcribed cDNA permits a very sensitive assay for viral specimens. Even though it is an absolute necessity to perform negative control reactions to exclude cross contamination and to perform assays at least in triplicate, the ability to detect small numbers of viral RNA molecules by PCR represents an extremely sensitive assay for HCV infection. The optimum region for PCR assay is the 5' leader RNA which is generally well–conserved among different HCV isolates. This excludes the possibility of missing viremia because of sequence heterogeneity. Since this 5' terminal region is also highly conserved within the 5' terminal region of pestiviral genomes (45% to 49%), perhaps PCR primers or hybridization probes in regions not highly conserved could be designed to avoid potential misdiagnosis. On the basis of a signature sequence for HCV (5' untranslated region), Cha *et al.*, (1991) have recently reported the development of a viral RNA assay using cDNA synthesis with reverse transcriptase, followed by PCR. This new assay was compared with the Ortho–Chiron C100–3 HCV enzyme–linked immunosorbent assay for antibodies to the C33c and C22 HCV antigens and the first reported set of HCV PCR primers designed from the NS3 domain. On the basis of the results obtained, the authors suggest that the use of probes and primers from the 5' PUT region (as opposed to primers from other regions) should not lead to false–negative results due to nucleic acid sequence variations in viral isolates.

Suboptimal specimen processing and storage conditions have been thought to give rise to false–negative results in the detection of HCV RNA in plasma or serum. In an effort to establish the influence of specimen handling in a serological laboratory on the rate of detection of HCV RNA by the cDNA polymerase chain reaction (cDNA–PCR), Cuypers *et al.*, (1992) tested routine serum samples and fresh–frozen samples from the same bleeding from confirmed anti–HCV–positive blood donors. When primers from the NS3/NS4 region were used, HCV RNA was detectable in fresh–frozen plasma from 67% of donors, whereas positive results were obtained with only 50% of the serum samples subjected to routine serological procedures. Analysis of the same samples with primers from the highly conserved 5' terminal region (5'–TR) revealed an HCV RNA detection rate of 92% for both routine and fresh–frozen samples. The yield of amplification product on routine samples was greatly reduced compared with that in the fresh–frozen plasma. Comparison of both primer tests for cDNA–PCR showed that the 5'–TR primer set was 10– to 100–fold more effective in detecting HCV RNA. A rapid decline in detectable HCV RNA of 3 to 4 log units was observed by these authors within 14 days when blood and serum were stored at room temperature. In contrast, no perceptible reduction in the cDNA–PCR signal was noticeable in freshly prepared serum stored at 4°C.

The fact that most PCR assays are qualitative limits their use to applications where only the presence or absence of target nucleic acid is to be determined. Thus, development of a calibrated external standard identical to target sequences or comparison with co-amplified, unrelated "reporter" RNA is necessary for quantitation of target sequences using the PCR technique. Recent studies (Lundeberg *et al.*, 1991) have reported the quantitation of target DNA/RNA by competitive PCR. Kaneko *et al.*, (1992) successfully utilized this competitive PCR to quantitate HCV RNA in sera of patients with chronic hepatitis. Quantitation of HCV RNA carries the potential for not only understanding the cause of HCV infection but also for evaluating the treatment for HCV infection.

A great deal of effort has been devoted to the detection of HCV RNA in clinical samples with a combination of reverse transcription and polymerase chain reaction (RT–PCR). A number of extremely sensitive PCR– based protocols are now known and have proved extremely useful in confirming and monitoring HCV infections (Ulrich *et al.*, 1990, Follett *et al.*, 1991; Sugitani *et al.*, 1992). However, as recently demonstrated by Zaaijer *et al.* (1993), the PCR used to detect HCV RNA should be interpreted with care as few laboratories currently undergo quality control testing for HCV PCR.

In order to increase the sensitivity and reliability of PCR analysis, a double–nested PCR (DN–PCR) has been exploited. In this method, the cDNA sample to be analyzed is first amplified with an outer set of primers and the amplified product from this primary PCR is subsequently reamplified in a separate reaction with a set of nested primers (Garson *et al.*, 1990). A pitfall of nested–set PCR appears to be that the second round PCR is set up in the presence of first–round PCR amplification products, increasing the risk of assay contamination by product DNA. As an alternative approach to nested–set PCR, a one–stage PCR amplification combined with Southern blot (Han *et al.*, 1991) liquid hybridization with radioactive oligonucleotide probes (Gretch *et al.*, 1992) has been used. Very recently, however, Gretch *et al.*, (1993) have described a one–stage PCR assay for HCV RNA which when combined with either liquid hybridization or Southern blot analysis has been shown to be equal in sensitivity to the nested–set PCR assay. Furthermore, this study also established that one round of PCR plus liquid hybridization offers sufficient sensitivity as well as less potential for contamination. The one step RNA PCR plus liquid hybridization offers a useful method for identification of those at risk of transmitting HCV infection and for monitoring responses to interferon therapy. The oligonucleotide probe specific for HCV nucleic acid besides increasing sensitivity also enhances the confidence level of positive results.

To circumvent the problem of contamination of secondary amplification with products accumulated from previous amplifications, a method for hemi–nested PCR (HN–PCR) using drop–in–drop–out primers has also been recently devised (Erlich *et al.*, 1991). This method can be accomplished in a single step. Romeo *et al.*, (1993) have utilized this novel reverse transcription polymerase chain reaction assay for the analysis of the prevalence of HCV complementary RNA in a set of 53 plasma specimens from blood donations that were repeatedly reactive for HCV antibodies with the first–generation enzyme immunoassay. Of 21 specimens that were also reactive for HCV antibodies by a four–antigen recombinant immunoblot assay (RIBA–II), 20 (95%) contained detectable levels of HCV RNA. Cryoprecipitate in three specimens reactive in the recombinant immunoblot assay led to apparent failure in detecting hepatitis C virus RNA but the authors confirm that repeat tests with redissolved cryoprecipitate subsequently revealed HCV RNA. With the future development of reliable automated RT–PCR assays, it appears that the detection of HCV RNA by RT–PCR may provide a practical confirmatory test of particular importance in the clinical management of apparently health blood donors who demonstrate HCV antibodies. The second generation four–antigen RIBA was found to be more specific than RIBA–I and to correlate with infectivity and detection of HCV RNA by PCR assay (Bryan *et al.*, 1993).

As well as providing valuable information concerning viremic status in the presence of anti–HCV and otherwise normal liver function, and as an efficient tool for monitoring therapeutic efficacy and diagnosis of HCV infection in chronic NANB hepatitis patients who may be seronegative, PCR assays have enabled the detection of vertical transmission of HCV from chronically infected mothers. The recent development of quantitative measurement methods for HCV RNA by competitive reverse transcription–PCR (Kaneko *et al.*, 1992; Kato *et al.*, 1993) may have broad application in the clinical study of hepatitis C virus infection.

V. CONCLUSION

Molecular genetics is an area of modern science which has attracted the attention of scientists worldwide, as has HCV, the major etiological agent of NANB hepatitis chronic liver disease and hepatocellular carcinoma (HCC). An effort has been made in this article to describe the effective use of molecular biological methods in the reliable diagnosis of hepatitis C disease.

Important inroads into the structure and function of the HCV have been made. The HCV virus is now known to contain a positive–strand RNA genome and the 5' terminal region is suggested to have an important regulatory role during viral replication. HCV has been projected as a novel type of virus. Nucleotide sequence similarity appears to be suggestive of HCV being closer in evolution to the pestiviruses than the flaviviruses. An apparent diversity in the nucleotide sequences of different HCV isolates has also been observed. Hepatitis C virus genotypes have been found to differ geographically with multiple genotypes encountered in certain areas.

The availability of sequence information as well as cDNA spanning the entire coding region has enabled the development of valuable diagnostic tools for HCV. The failure to detect circulating antibodies to the C–100 antigen has been observed with post–transfusion NANB hepatitis patients and false–positives also reported. The use of additional recombinant–derived viral proteins in the development of an ultra–sensitive assay for anti–HCV with quicker detection of seroconversion will undoubtedly be necessary.

In view of the low titer of circulating virus, a common occurrence in HCV infection, extensive efforts have been directed towards the detection of HCV RNA in clinical samples. The application of PCR techniques with amplification of reverse– transcribed cDNA has provided an effective means for assaying HCV infections. A number of highly sensitive PCR–based assays (RT–PCR) are now in use. In order to increase the sensitivity and reliability of PCR analysis, a double–nested PCR (DN–PCR) has been exploited but recent reports indicate the risk of assay contamination by product DNA. A very recent one–stage PCR assay for HCV RNA combined with either liquid hybridization or Southern blot analysis equal in sensitivity to the nested–set PCR assay, but with reduced potential for contamination, shows promise. Further development of reliable and automated RT–PCR assays appears to be of particular interest.

REFERENCES

Alberti, A. (1991). *J. Hepatol.* **12,** 279–282.

Alter, H. J., Purcell, R. H., Shih J. W., Melpolder, J. C., Houghton, M., Choo, Q. L. and Kuo, G. (1989). *New Engl. J. Med.* **321,** 1494–1500.

Boudart, D., Lucas, J. C., Adjou, C. and Muller, J. Y. (1992). *Lancet* **338,** 372.

Bruix, J., Barrera, J. M., Calvet, X., Ercella, G., Costa, J., Sanchez–Tapias, J. M. and Ventura, M. (1989). *Lancet* **2,** 1004–1005.

Bryan, S. P., Sjogren, M. H., Malone, S. L., MacArthy, P., Kao, T. C., Wagner, K., Sheffield, S., Smith, E. and Perine, P. L. (1993). *J. Infect. Dis.* **167,** 715–719.

Bukh, J., Purcell, R. and Miller, R. (1993). *Proc. Natl. Acad. Sci.* **90,** 8234–8238.

Cha, T. A., Kolberg, J., Irvine, B., Stempien, M., Beall, E., Yano, M., Choo, Q. L., Houghton, M., Kuo, G., Han, J. H. and Urdea, M.S. (1991). *J. Clin. Microbiol.* **29,** 2528–2534.

Choo, Q. L., Richman, K. H., Han, J. H., Berger, K., Lee, C., Dong, C., Gallegos, C., Cott, D., Medina–Selby, A., Barr, D. J., Weiner, A. J., Bradley, D. W., Duo, G. and Houghton, M. (1991). *Proc. Natl. Acad. Sci. USA* **88,** 2451–2455.

Choo, Q. L., Kuo, G., Weiner, A. J., Overby, L. R., Bradley, D. W. and Houghton, M. (1989). *Science* **244,** 359–362.

Choo, Q. L., Weiner, A. J., Overby, L. R., Kuo, G., Houghton, M. and Bradley, D. W. (1990). *Br. Med. Bull.* **46,** 432–441 .

Collett, M. S., Anderson, D. K. and Retzel, E. (1988). *J. Gen. Virol.* **69,** 2637–2643.

Collett, M. S., Moenning, V. and Horzinek, M. C. (1989). *J. Gen. Virol.* **70,** 253–266.

Collett, M. S., Wiskerchen, M. A., Welniak, E. and Belzer, S. K. (1991). *Arch. Virology* **3,** 19–27.

Colombo, M., Kuo, G., Choo, Q. L., Donato, M. F., Del Ninno E., Tommasim M. A. and Droguardi, N. (1989). *Lancet* **2,** 798–799.

Cuypers, H. T. M., Bresters, D., Winkel, I. N., Reesink, H. W., Weiner, A. J., Houghton, M., Van der Pael, C. L. and Leslie, P. N. (1992). *J. Clin. Microbiol.* **30,** 3220–3224.

Donahue, J. G., Munz, A., Ness, P. M., Brown, D. E., Yawm, D. H., McAIlister, H. A. and Reitz, B. A. (1992). *New Engl. J. Med.* **327,** 369–373.

Erlich, H. A., Gelfand, D. and Sninsky, J. J. (1991). *Science* **252,** 1643–1651.

Esteban, J. L., Esteban, R., Viladoumiu, L., Lopez–Talavera, J. C., Gonzalez, A., Hernandez, J. M. and Roget, M. (1989). *Lancet* **2,** 294–297.

Esteban, J. L., Gonzalez, A., Hernandez, J. M., Viladomiu, I., Sanchez, C., Lopez–Talavira, J. C. and Lucea, D. (1990). *New Engl. J. Med.* **323,** 1107–1112.

Follett, E. A. C., Dow, B. C., McUnish, F., Yap, P. L., Hughes, O., Michell, R. and Simmonds, P. (1991). *Lancet* **338,** 1024.

Garson, J. A., Tedder, R. S., Briggs, M., Take, P., Glazebrook, J. A., Trute, A., Paraker, D., Barbara, J. A. J. and Contreras, M. (1990). *Lancet* **335,** 1419–1422.

Gretch, D. R., Lee, W. and Corey, L. (1992). *J. Clin. Microbiol.* **30,** 2145–2149.

Gretch, D. R., Wilson, J. J., Carithers, R. L., Corazon, J. R., Han, J. H. and Corey, L. (1993). *J. Clin. Microbiol.* **31,** 289–291.

Han, J. H., Shyamala, V., Richman, K. H., Brauer, M. J., lrvine, B., Undea, M. S. and Tekamp–Olson, P. (1991). *Proc. Natl. Acad. Sci, USA* **88,** 1711–1715.

He, L. F., Alling, D., Popkin, T., Shapiro,M., Alter, H. J. and Purcell, R. H. (1987). *J. Infect. Dis.* 156, 636–640.

Hijikata, M., Kato, N. and Ootsuyama, Y. (1991). *Biochem. Biophys. Res. Commun.* **175,** 220–228.

Kaneko, S., Murakami, S., Unoura, M. and Kobayashi, K. (1992). *J. Med. Virol.* **37,** 278–282.

Katayama, T., Kikuchi, S., Tanaka, Y., Saito, I., Miyamura, T., Choo, Q. L. and Houghton, M. (1990). *Transfusion* **30,** 374–376.

Kato, N., Hijikata, M., Ootsuyama, Y., Nakagawa, M., Ohkoshi, S., Sugimura, T. and Shimotomo, K. (1990). *Proc. Natl. Acad. Sci. USA* **87,** 9524–9528.

Kato, N., Yokosuka, O., Hosoda, K., Ito, Y., Ohto, M. and Omata, M. (1993). *Hepatology* **18,** 16–20.

Kubo, Y., Takeuchi, K., Boonmar, S., Katayama, T., Choo, Q. L., Kuo, G., Weiner, A. J., Bradley, D. W., Houghton M., Saito I. and Miyamura T. (1989). *Nucleic Acids Res.* **17,** 103ti7–10372.

Kuo, G., Choo, Q. L., Alter, H. J., Getnick, G. I., Redeker, A. G., Purcell, R. H., Miyamura, T., Mitamura, J. L., Alter, M. J., Steven, C. E., Tegtmeier, G. E., Bonino, F., Colombo, M., Lee, W. S., Kuo, C., Berger, K., Shuster, J. R., Overby, L. R., Bradley, D. W. and Houghton, M. (1989). *Science* **244,** 362–364.

Kurstak, E. (1993). *Viral hepatitis: Current status and issues*. Springer–Verlag, Wien, New York, 1–217.

Lundeberg, J., Wahlberg, J. and Uhlen, M. (1991). BioTechniques **10,** 68–75.

Machida, A., Ohnuma, H., Tsuda, F., Munekata, E., Tanaka, T., Akahane, Y., Okamoto, H. and Mishiro, S. (1992). *Hepatology* **16,** 886–891.

MacFarlane, I. G., Smith, H. M., Johnson, P. J., Bray, G. P., Vergani, D. and Williams, R. (1990). *Lancet* **335,** 754–757.

Miller, R. H. and Purcell, R. N. *Proc. Natl. Acad. Sci. USA* 1990, **87,** 2057–2061.

Nishioka, K., Watanabe, J., Furuta, S., Tanaka, E., Lino, S., Suzuki, H. and Tsuji, T. (1991). *Cancer* **67,** 429–433.

Okamoto, H., Okada, S., Sugiyama, Y., Yotsumoto, S., Tanaka, T., Yoshizawa, H., Tsuda, F., Miyakawa, Y. and Mayami, M. (1991). *Jap. J. Exptl. Med.* **60,** 167–177.

Okamoto, H., Tsuda, F., Machida, A., Munikata, E., Akahne, Y., Sugai, Y., Mashiko, K. and Mayami,A. (1992). *Hepatology* **15,** 180–186.

Purcell, R. H. (1993). In *20th Anniversary of Ausria Symposium* pp. 29–30. Abbott, Chicago.

Rice, C. M., Strauss, E. G. and Strauss, J. H. (1986). In *Togaviridae and Flaviviridae.* (S. Schlesinger and M. Schelsinger, eds.) pp. 279–326. Plenum Press, New York.

Romeo, J. M., Ulrich, P. L., Busch, M. P. and Vyas, G. N. (1993). *Hepatology* **17,** 188–195.

Simmonds, P., Zhang, L. Q., Watson, N. G., Rebus, S., Ferguson, E. D., Bulfe, P. and Leadbetter, G. N. (1990). *Lancet* **336,** 1469–1472.

Skidmore, S. (1990). *Lancet* **335,** 1346.

Stark, R., Rumenapf, T., Meyers, G. and Thiel, H. J. (1990). *Virology* **174,** 286–289.

Sugitani, M., Inchauspe, G., Shindo, M. and Prince, A. M. (1992). *Lancet* **339,** 1018–1019.

Takada, N., Takasi, S., Takada, A. and Date, T. (1992). *Lancet* **339,** 808.

Takamizawa, A., Mori, C., Fuke, I., Manabe, S., Murakami, S., Fujita, J., Oneshi, E., Andoh, T., Yoshida, I. and Okayama, H. (1991). *J. Virol.* **65,** 1105–1113.

Takeuchi, K., Boonmar ,S., Kubo, Y., Katoyama, T., Harada, H., Ohbayashi, A. and Choo, Q. L. (1990a). *Gene* **91,** 287–291.

Takeuchi, K., Kubo, Y., Boonmar, S., Watanabe, Y., Katayama, T., Choo, Q. L. and Kuo, G. (1990b). *Nucleic Acids Res.* **18,** 4626.

Tanaka, E., Kiyosawa, K., Sodeyama, T., Nakamo, Y., Yoshizawa, K., Hayata, T. and Schimiza, S. (1991). *J. Med. Virol.* **33,** 117–121.

Ulrich, P. P., Romeo, J. M., Lane, P. K., Kelly, I., Daniel, L. J. and Vyas, G. N. (1990). *J. Clin. Invest.* **86,** 1609–1614.

Van der Poel, C. L., Reesink, H. W., Lelie, P. N., Leentvaar–Kuypers, A., Choo, Q. L., Kuo, G. and Houghton M. (1989). *Lancet* **2,** 297–298.

Van der Poel, C. L., Cuypers, H. T. M. and Reesink, H. W. (1991). *Lancet* **337,** 317–319.

Weiner, A. J., Kuo, G., Bradley, D. W., Bonino, F., Saracco, G., Lee, C. and Rosenblatt,, J. (1990). *Lancet* **335,** 1–3.

Weiner, A. J., Brauer, M. J. andRosenblatt, J. (1991). *Virology* **180,** 842–848.

Zaaijer, H., Cuypers, H., Reesink, H., Winkel, I., Gerken, G. and Lelie, P. (1993). *Lancet* **341,** 722–724.

5

Synthetic Peptides in Virology

The Use of Synthetic Peptides in Retrovirus Diagnosis

A. Närvänen, M. Korkolainen, S. Kontio,
M–L. Huhtala, X. Imai and M. Hayami

I. INTRODUCTION

The first great milestone in the history of synthetic peptides as tools in the biological sciences was achieved at the beginning of 1960's, when Merrifield published the method of solid phase peptide synthesis (Merrifield, 1963). Although, peptides had been synthetized earlier, the new technology made it possible to synthetize peptides in milligram quantities during few weeks instead of several months. On the other hand the revolution in DNA technology during the decade of 1970 increased dramatically and is increasing continuously the data base of amino acid sequences of biologically important proteins. However, limitations on the systematic use of synthetic peptides to identify all possible sequential epitopes to a given protein sequence remained until the 1980s. Several attempts to rely on guess–work or predictive algorithms to determine, for instance, the loci of antigenic epitopes in viral proteins have been published in the literature (Pfaff *et al.*, 1982; Tainer *et al.*, 1984; Hopp *et al.*, 1986; Thornton *et al.*, 1986). However, the reliability of these predictive algorithms for sequential antigenic epitopes has been shown to be not better than by chance alone (Westhof *et al.*, 1984; Getzoff *et al.*, 1987, 1988). The first and thus far the most effective and practical procedure allowing the systematic synthesis of all possible sequential epitopes is the so called "Multipin Peptide Synthesis System" developed by Geysen *et al.* (1984). These developments have extensively increased the use of synthetic peptides in the fields of immunology and vaccine research during the last few years.

A. Närvänen, M. Korkolainen, S. Kontio and M–L. Huhtala • Labsystems Research Laboratories Oy, SF–0080, Helsinki, Finland. *X. Imai and M. Hayami* • Institute for Virus Research, Kyoto University, 606 Kyoto, Japan.

Applied Virology Research, Vol. 3, Edited by
E. Kurstak *et al.*, Plenum Press, New York, 1994

II. MULTIPIN PEPTIDE SYNTHESIS SYSTEM

The pin system for peptide synthesis provides large numbers of peptides at a reasonable cost, and time as compared with alternative methods of synthesis (Geysen *et al.*, 1984; Geysen *et al.*, 1987a). The principle of this technique is to synthetize partially overlapping peptides, which cover the whole studied protein sequences. Peptides are synthetized manually on polyethylene pins using the basic technique of Merrifields solid phase synthesis (1963). One set of polyethylene pins has been designed to fit into the most commonly used 96 microtitration immunoassay well format. Kits containing polyethylene pins with an already existing covalently coupled t–butyloxycarbonyl protected beta–alanine, to complete a peptide like spacer, are commercially available (Cambridge Research Biochemicals, Cambridge, U.K.). The strategy allows the synthesis of several hundreds of different peptides simultaneously (Geysen *et al.*, 1987a). The total time of the synthesis depends on the length of the synthetized peptides meaning that each amino acid addition requires one day. The method is simple and no extensive knowledge of organic chemistry is required. The pin system provides an important tool in viral immunolgy from diagnostic test to viral vaccine design. We will describe briefly the methods of its application in virology.

A. B– and T–cell Epitope Mapping

In the original approach of Geysen (1984), hexapeptides were selected for the B–cell epitope mapping. This was based on the determination of all antigenic sites of myoglobin (Atassi *et al.*, 1980, 1984). Recently, the synthesis has been performed at the octapeptide level, since it has been shown that whilst about 90% of linear epitopes are 6 amino acids or less, approximately 10% require 7 and <1% 8 amino acids (Geysen *et al.*, 1988). However, the octapeptide level is not an upper limit of the length of the peptide in this method, but can be increased according to the requirements. The immunoreactivity of the synthetized peptides, while still covalently bound on the pins, can be tested to a given antibody with conventional immunochemical methods (enzyme immunoassay, radio immuno–assay, fluorecence immunoassay and lumino immunoassay). Since synthetic peptides are chemically very stable, they can be reused after removing the bound antibodies from the pins. The pin bound peptide approach is ideal for identification of all sequential B–cell epitopes from viral proteins (Geysen *et al.*, 1984; Geysen *et al.*, 1985; Meloen *et al.*, 1986, 1987; Geysen *et al.*, 1987a, b; Goudsmit *et al.*, 1988a, b, 1989; Cason *et al.*, 1989; Papsidero *et al.*, 1989; Ralston *et al.*, 1989; Shugla, D.D:, 1989; Singh *et al.*, 1989; Dillner, 1990; Fridell *et al.*, 1990; Gombert *et al.*, 1990; Lauriella *et al.*, 1990; Lozzi *et al.*, 1990; Stemler *et al.*, 1990; Varnier *et al.*, 1990; Närvänen *et al.*, 1991; Roivainen *et al.*, 1991). However, the mapping of T–cell epitopes requires liquid–phase peptides. In addition, cell–cultures require high standard of purity and non–toxicity of the synthetized peptides. To fulfill these requirements for T–cell cultures a diketopiperazine

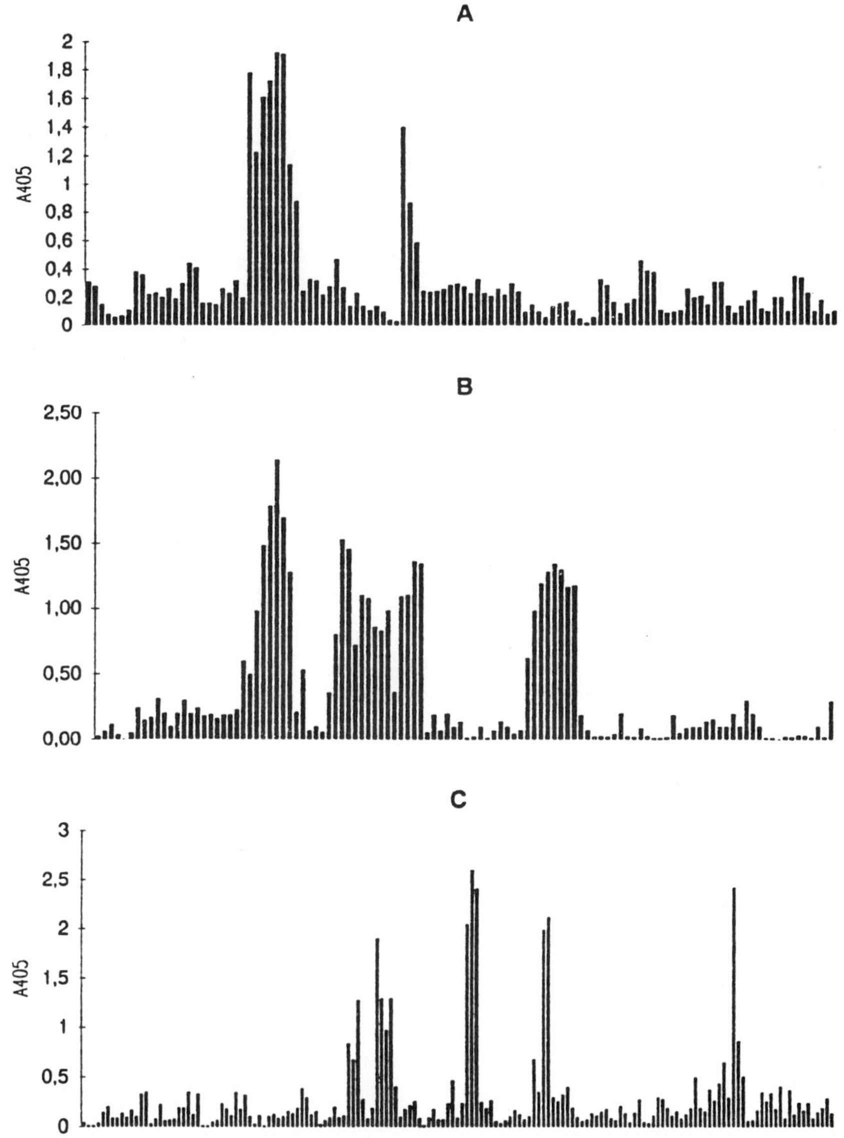

Figure 1

The epitope mapping profiles of the envelope proteins of three human retroviruses. Every bar represents an individual peptide and the height of the bar the reactivity in indirect EIA. The epitope mapping profile of: A) the transmembrane protein gp41 of HIV–1 with HIV–1 antibody positive human sera; B) the transmembrane protein gp36 of HIV–2 with HIV–2 antibody positive human sera; and C) the envelope glycoproteins of HTLV–I with HTLV I antibody positive human sera.

(DKP) method has been developed to cleave the synthetized peptides gently out of the pins (Maeji *et al.*, 1990). Peptides are cleaved into 96 microtitration well format at neutral or alkaline pH for example, into phosphate buffered saline (PBS) or bicarbonate buffer. The DKP–cleaved peptides have been shown to be highly purified without cell–toxicity and have been used in the T–cell cultures without any additional purification steps (Van der Zee *et al.*, 1989; Maeji *et al.*, 1990, Bray *et al.*, 1990).

B. Determining the Length of an Epitope

It is usual to find that almost all epitopes occur in the clusters of overlapping peptides in epitope mapping of viral proteins (Geysen *et al.*, 1987a). This suggests that the length of an epitope is shorter than the peptides, i.e., octapeptides, which are used in original mapping. The accurate length of the found epitopes can be determined by synthetizing overlapping peptides of different length covering the original epitope area (Schaaper *et al.*, 1990). The range may vary, i.e., starting from tetrapeptides and ending to heptapeptides. The exact determination of the length of the epitope has an advantage in lowering the number of the peptides in the studies of the epitope structure at the single amino acid level.

C. Epitope Structure Determination at a Single Amino Acid Level

The reaction of an IgG molecule and an epitope is based on certain side–chains of the amino acids in an epitope. Some of the amino acid side–chains are responsible in direct interaction, i.e., are the contact residues while some amino acids are not in direct contact but are needed as building blocks of an epitope (Amit *et al.*, 1986; Laver, W.G., 1990). These amino acid side chains are not involved in an energetically favorable interaction with the antibody. During the determination of the contact residues and building blocks all the amino acids are replaced to 19 other amino acids. The reactivity of an antibody with all peptide–variants determines the essential and non– essential amino acid side–chains for the direct reaction with IgG–molecule (Schoofs *et al.*, 1988). In addition, amino acids can be replaced with un–natural amino acids (D–amino acids, etc.). This strategy allows the design of artificial epitopes and helps to understand the rules for antibody binding in terms of both residues constraints and structural requirements (Geysen *et al.*, 1987b).

D. Mimotope Design

The limitation of the conventional epitope mapping method is the requirement of the continuity of an epitope. An alternative approach to mimick discontinuous or assembled epitopes (Berzofsky *et al.*, 1982; Benjamin *et al.*, 1983) is to apply the pin system and start with an antibody instead of an antigen.

Monclonal antibody can be used to design the complementary structure to it itself. Two different practical approaches have been described for the design of an antigen to a specific antibody: A) In the combinatorial strategy the synthesis is started from grouped pools of mixtures of peptides. The final binding peptide is progressively "decoded"; and B) In the incremental strategy the epitope is progressively "built up" starting with a dipeptide, and extending via selection until the appropriate degree of complementarity is achieved. A common approach in this strategy is the parallel use of both natural L–amino acids and "un–natural" D–amino acids. The designed epitope do not have to resemble the original epitope structure against which antibody was elicited but may be totally artificial structure without any model in the antigen molecule (Geysen *et al.*, 1986 a, b).

III. CHARACTERIZATION OF SEROLOGICALLY IMPORTANT EPITOPES OF HUMAN RETROVIRAL PROTEINS

Human pathogenic retroviruses consist of two groups of viruses: the first group consisting of human T–cell leukemia viruses type I and II (HTLV–I and HTLV–II) and the second group human immunodeficiency viruses type 1 and 2 (HIV–1 and HIV–2). These four viruses are structurally and serologically closely related. The average homology between HIV type 1 and 2 proteins being 40% (Clavel *et al.*, 1986). The cross–reactive rate between HIV type 1 and 2 proteins varies from 20 to 80% . The clinical syndromes associated with the two retrovirus groups are quite dissimilar; the HTLV–group results in T–cell leukemia whereas HIV–group cause deficiency of T–cell population. For better understanding of the epidemiology and clinical features of these infections it is important to distiguish between the causative agents.

The first commercially available products for the detection of antibodies to HIV and HTLV were EIA kits based on antigens obtained from disrubted virions grown in vitro in continuous human T–cell leukemia lines. However, the use of synthetic peptide based antigens in HIV and HTLV EIAs is increasing continuosly due to the better characteristics of these antigens compared to viral lysate or recombinant based antigens. Chemically synthetized peptides representing conserved immunodominant epitopes of retroviral proteins have been shown to provide excellent tools for the detection of retroviral antibodies in many respects; like typing or simultaneous detection of the different retroviral infections (Wang *et al.*, 1986; Rosen *et al.*, 1987; Gnann *et al.*, 1987 a and b; Norrby *et al.*, 1987, 1989; Huhtala *et al.*, 1987; Kemp *et al.*, 1988; NÄrvÄnen *et al.*, 1988; Korkolainen *et al.*, 1990; Kuroda *et al.*, 1990; Lal *et al.*, 1990, 1991; Varnier *et al.*, 1991).

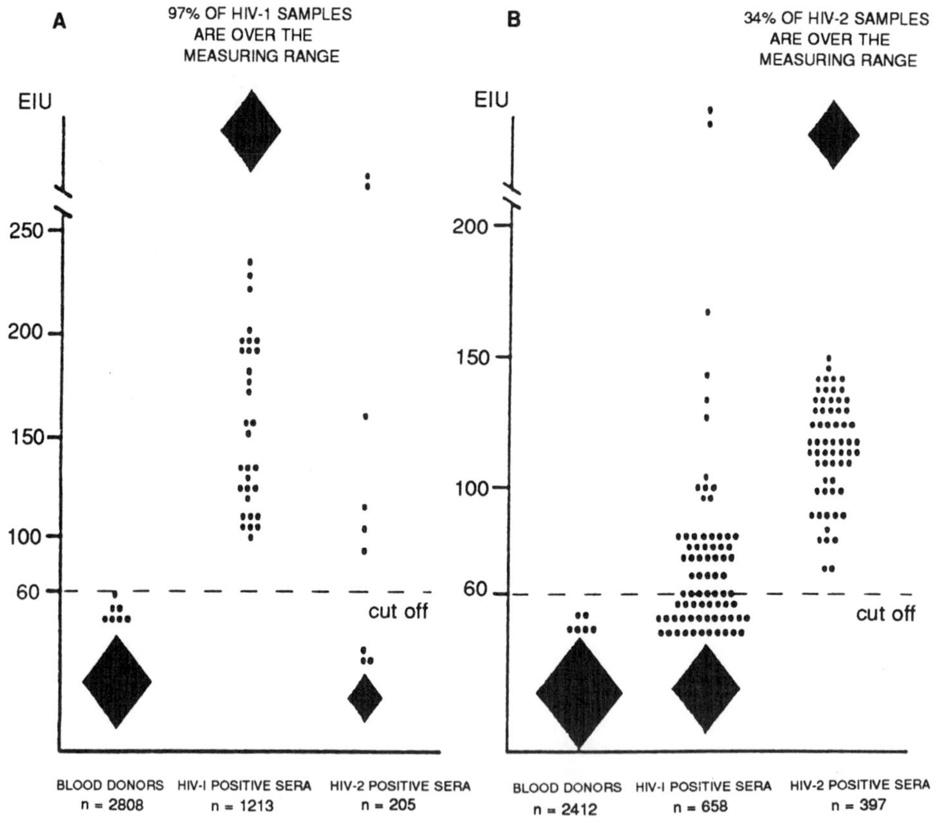

Figure 2

The reactivity of: A) the HIV–1 type–specific EIA with HIV antibody negative Finnish blood donors sera, HIV–1 and HIV–2 antibody positive human sera.; B) the HIV–2 type–specific EIA with HIV antibody negative Finnish blood donors sera, HIV–1 and HIV–2 antibody positive human sera. The results are calculated as enzyme immuno units (EIU) according to the equation:

$$EIU = \frac{\text{abs (S)} - \text{abs (B)}}{\text{abs (PC)} - \text{abs (B)}} \times 100$$

where abs (S) = absorbance value of the sample, abs (B) = absorbance value of blank and abs (PC) = absorbance value of the positive control. The cut off is adjusted to be 60 EIUs in both tests.

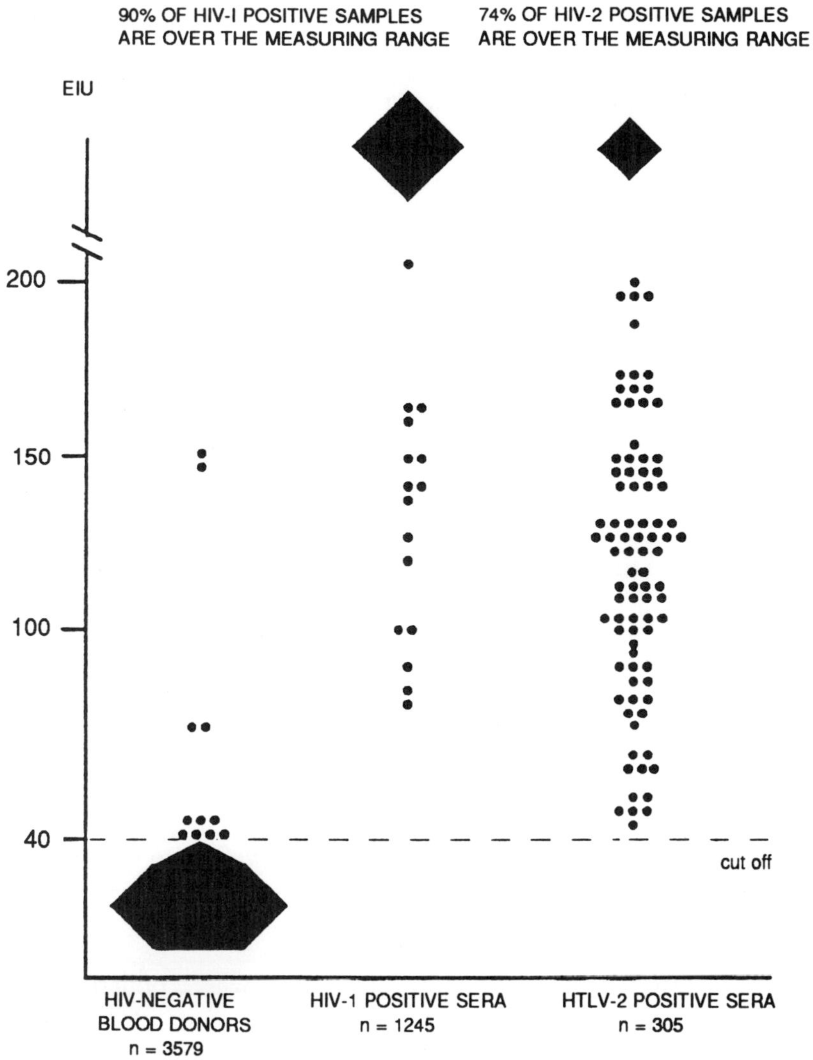

Figure 3

The reactivity of the combined HIV–1 and HIV–2 specific EIA with HIV antibody negative, HIV–1 and HIV–2 antibody positive human sera. The cut off is adjusted to be 40 EIUs (see text in Fig. 2).

A. Epitope Mapping of Retroviral Proteins

By using the "Multipin Peptide Synthesis System", we have studied the antigenic structure of the transmebrane proteins gp41 of HIV–1 and gp36 of HIV–2 as well as the envelope glycoprotein of HTLV I. The protein sequences were divided into 14 amino acid windows. The window was moved along the sequence in three amino acid steps starting from the amino terminal end of the protein and ending at the COOH–terminal end of the protein. This 14 amino acid approach was selected, since the number of synthesized peptides was lowered to one third of the original strategy while every octapeptide was synthetized twice or three times. After synthesis was completed the 14 amino acid length peptides were tested with HIV–1, HIV–2 and HTLV I antibody positive and negative human sera using EIA method as described (Roivainen *et al.*, 1991). In Figure 1 has been shown typical scanning profiles of HIV–1 and HIV–2 transmebrane proteins and HTLV–I envelope proteins with the corresponding HIV–1, HIV–2 and HTLV–I antibody positive human sera.

B. EIA Applications

After the identification of the serologically important antigenic epitopes, synthetic peptides, derived from these regions, were used as antigens in convetional indirect EIAs (Huhtala *et al.*, 1987; Närvänen *et al.*, 1988; Närvänen, 1990; Varnier *et al.*, 1991). The most type–specific peptides from HIV–1 and HIV–2 transmembrane proteins were applied into the separate EIAs and tested with the panels of HIV–1 positive (n = 1213) human sera from Finland and Italy and HIV–2 positive (n = 205) human sera from West–Africa. The HIV antibody negative (n = 2080) human sera were from Finnish blood donors (Figure 2). The cross–reactive rate in the HIV–1 type–specific EIA with HIV–2 positive human sera was 2.9%, whereas in the HIV–2 type–specific test the cross–reactivity with HIV–1 positive human sera was 5.8%. For screening purposes the most immunodominant epitopes both from HIV–1 and HIV–2 were used in the same EIA. The combination of the peptides detected all tested HIV positive human sera ((n = 1550; Figure 3) the specificity being 99.7 % with the Finnish blood donation material (n = 3579) .

The most immunoreactive peptides of HTLV–I were applied in EIA and tested with HTLV–I positive human sera from Japan, where HTLV I occurs as endemic. The HTLV–I antibody positivity of the sera was based on immunofluorecence testing. The cross–reactive rate of the HTLV–I test was 2% with the cohort of HIV–1 positive human sera (Figure 4). The specificity of the EIA was 100 % with Finnish blood donation sera (n = 116).

These results indicate that synthetic peptides can give clear serological differentiation between structurally closely related retroviruses thus providing ideal tools for epidemiological studies and for supporting clinical features of the corresponding infections. It should be noted that the cross– reactive rate in viral based tests between HIV–1 and HIV–2 varies from 40 to 80 %. In addition, it is

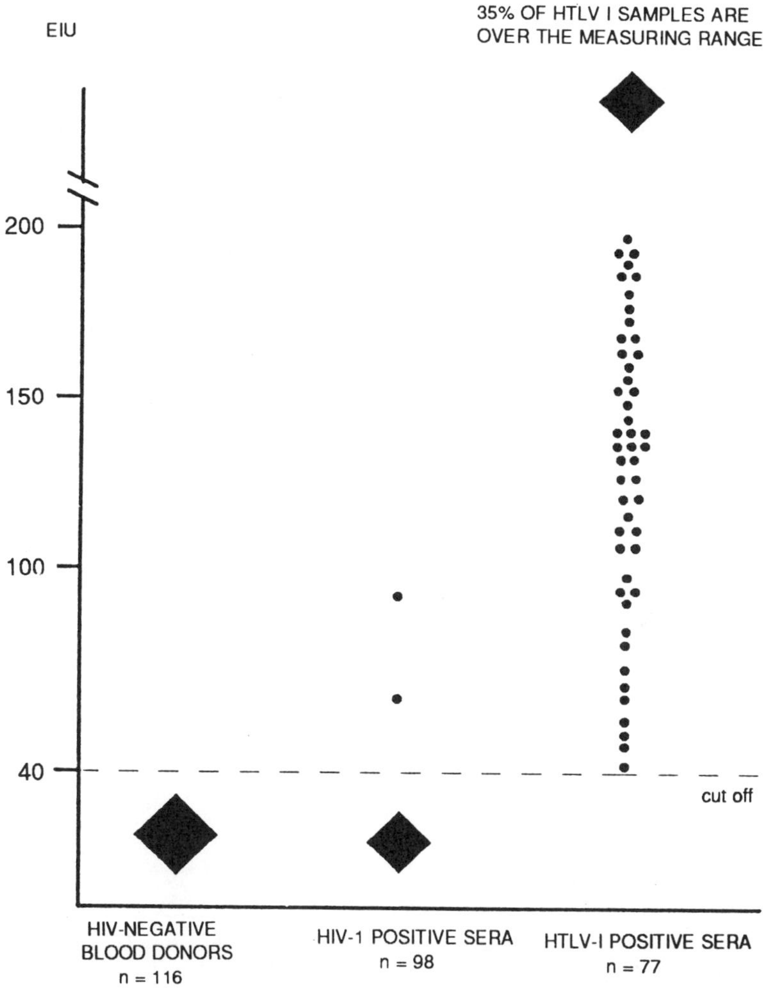

Figure 4

The reactivity of the HTLV I specific EIA with HIV antibody negative, HIV–1 and HTLV I antibody positive human sera. The cut off is adjusted to be 40 EIUs (see text for Fig. 2).

easy to combine different synthetic peptides in the same test thus allowing the detection of different viral infections simultaneously.

IV. THE FUTURE OF SYNTHETIC PEPTIDES IN VIRAL IMMUNOLOGY

The fast developments in DNA and synthetic peptide technologies is now a challenge to resolve the antigenic structures of the most important and pathogenic viruses. The mapping of the B–cell and T–cell determinants from viral proteins has already lead to immunologically important basic findings and serologically important applications. The T–cell immunity research using synthetic peptides is now increasing. At least one vaccine that is based entirely on peptides has been evaluated (DiMarchi *et al.*, 1986), but the mapping of natural T–cell epitopes as well as the building of artificial mimotopes may lead to the totally new strategies in vaccine design (McGregor *et al.*, 1990).

REFERENCES

Amit, A. G., Mariuzza, R. A., Phillips, S. E. V., and Poljak, R. J. (1986). *Science* **233,** 747–753.

Atassi, M. Z. (1980). *Mol. Cellular Biochem.*, **32,** 21–43.

Atassi , M. Z. (1984). *Europ. J. Biochem.* **145,** 1–20.

Benjamin, D. C., Daigle, L. A., and Riley, R. L. (1983). In *Protein Conformation as an Immunological Signal* (F. Gelada, V. N. Shumaker, E. E. Sercarz. eds.), Plenum Press, New York.

Berzofsky, J. A., Buckenmeyer, G. K., Hicks, G., Grud, F. R. N., Feldman, R. J., and Minna, J. (1982). *J. Biol. Chem.* **257,** 3189.

Bray, A. M., Maeji, N. J., and Geysen, H. M. (1990). *Tetrahedron Letters* **31,** 5811–5814.

Cason, J., Patel, D., Naylor, J., Lunney, D., Shepherd, P. S., Best, J. M., and McCance, D. J. (1989). *J. Gen. Virol.* **70,** 2973.

Clavel, F., Guyader, M., Guétard, D., Sallé, M., Montagnier, L., and Alizon, M. (1986). *Nature* **324,** 691–695.

Gnann Jr, J. W., Nelson, J. A., and Oldstone, M.B.A. (1987a). *J.Virol.* **61,** 2639–2641.

Gnann Jr, J. W., McCormic, J. B., Mitchell, S., Nelson, J. A., and Oldstone, M. B. A. (1987b). *Science* **237,** 1346–1349.

Dillner, J. (1990). *Intl. J. Cancer* **46,** 4, 703–711.

DiMarchi, R., Brooke, G., Gale, C., Cracknell, V., Doel, T., and Mowat, N. (1986). *Science*, **232,** 639–641.

Fridell, E., Blinkovsky, A., Mannervik, M., Trojnar, J., and Wahren, B. (1990). *Proceedings of the 21st European Peptide Symposium,* Barcelona.

Getzoff, E. D., Geysen, H. M., Rodda, S. J., Alexander, H., Tainer, J. A. and Lerner, R. A. (1987). *Science* **235,** 1191–1196.

Getzoff, E. D., Tainer, J. A., Lerner, R. A., and Geysen, H. M. (1988). *Adv. Immunol.* **43**, 1–98.
Geysen, H. M., Meloen, R. H., and Barteling, S. J. (1984). *Proc. Natl. Acad. Sci. USA* **81**, 3998–4008.
Geysen, H. M. (1985) *Immunology Today* **6**, 12.
Geysen, H. M., Rodda, S. J., and Mason, T. J. (1986a). *Mol. Immunol.* **23**, 709–715.
Geysen, H. M., Rodda, S. J., and Mason, T. J. (1986b). *Synthetic Peptides as Antigens.* Ciba Fndn. Symp. **119**, 130.
Geysen, H. M., Rodda, S. J., Mason, T. J., Tribbick, G., and Schoofs, P.G. (1987a). *J. Immunol. Meth.* **102**, 259–274.
Geysen, H. M., Tainer, J. A., Rodda, S. J., Mason, T. J., Alexander, H., Getzoff, E. D., and Lerner, R. A. (1987b). *Science* **235**, 1184.
Geysen, H. M., Mason, T. J., and Rodda, S. J. (1988). *J. Molec. Recognition* **1**, 32–41.
Gombert, F. O., Schaude, R., Troger, W., Jung, G., Rubsamen– Waigmann, H., Blecha, W., Pfeifer, S., Krohn, K., Tahtinen, M., Ovod, V., Ranki, A., Shi, L., and Wernet, P. (1990). *Proc. 21st Europ Peptide Symp.*, Barcelona.
Goudsmit, J., Debouk, C., Meloen, R. H., Smit, L., Bakker, M., Asher, D. M., Wolff, A. V., Gibbs, C. J., and Gajdusek, D. E. (1988a). *Proc. Natl. Acad. Sci. USA.* **85**, 4478.
Goudsmit, J., Boucher, C. A. B., Meloen, R. H., Epstein, L. G., Smit, L. A., Van der Hoeck, L., and Bakker, M. (1988b). *J. AIDS* **2**, 157.
Goudsmit, J., Meloen, R. H., Brasseur, R., and Barin, F. (1989). *Ibid.* **2**, 297.
Hopp, T. (1986). *J. Immunol. Meth.* **88**, 1–18.
Huhtala, M.–L., Närvänen, A., Korkolainen, M., Kontio, S., Partanen, P., Suni, J., and Vaheri, A. (1987). In *Rapid Methods and Automation in Microbiology and Immunology.* (A. Barlows, R. C. Tilton, and A. Turano, eds.),Brescia Academic Press, Brescia.
Kemp, B. E., Rylatt, D. B., Bundsen, P. G., Doherty, R. R., McPhee, D. A., Stapelton, D., Cottis, L. E., Wilson, K., John, M. A., Kjan, J.M., Dihn, D.P., Miles, S., and Hillyard, C. J. (1988). *Science* **241**, 1352–1354.
Korkolainen, M., Närvänen, A., Varnier, O., Lillo, F., Kontio, S., Rescaldani, R., Visconti, A., and Huhtala, M–L. (1990). In: *Innovation and Perspectives in Solid Phase Synthesis– Peptides, Polypeptides and Oligonucleotides: Macro–Organic Reagent and Catalyst.* (R. Epton, ed.). SPCC (UK) Ltd., West Midland, England
Kuroda, N., Washitani, Y., Shiraki, H., Kiyokawa, H., Ohno, M., Sato, H., and Maeda, Y. (1990). *Intl. J. Cancer.* **45**,865–868.
Lal, R. B., Rudolph, D. L., Lairmore, M. D., Khabbaz, R. F., Garfield, M., Coligan, J. E., and Folks, T. M. (1990). *J. Inf. Dis.* **163**, 41–46.
Lal, R. B., Rudolph, D. L., Griffis, K. P., Kitamura, K., Honda, M., Coligan, J. E., and Folks, T. M. (1991). *J. Virol.* **65**,1870– 1876.
Lauricella, R. J., Shukla, D. D., Plompen, S., Morgan, P., Stanton, D., Harrison, M., and McKerral, M. (1990). *Proc. 15th Ann. Conf. Protein Struct. Funct.*, Lorne, Australia.
Laver, W., Air., G. M., Webster, R. G., and Smith–Gill, S. J. (1990). *Cell* **61**, 553–556.

Lozzi, L., Bracci, L., Corti, M., Cusi, M. G., Santucci, A., Valencin, P. E. and Neri, P. (1990). *Proc. 21st Europ. Peptide Symp.*, Barcelona.

Maeji, N. Bray, A. M., and Geysen, H. M. (1990). *J. Immunol.* Meth. **134**, 23–33.

McGregor, D., Triantafyllou, B., Tribbick, G., and Geysen, H. M. (1990). *Proc. 15th Ann. Conf. Protein Struct. Funct.*, Lorne, Australia.

Meloen, R. H., and Barteling, S. J. (1986). *Virology* **149**, 55.

Meloen, R. H., Puyk, W. C., Meiger, D. J .A., Lankhof, H., Posthumus, W. P. A., and Schaaper, W. M. M. (1987). *J. Gen. Virol.* **68**, 305.

Merrifield, B. (1963). *J. Am. Chem. Soc.* **85**,2149–2152.

Norrby, E., Biberfeld, G., Chiodi, F., von Gegerfeld, A., Nauclér, A., Parks, D. E., and Lerner, R. (1987). *Nature* **329**, 248–250.

Norrby, E., Parks, D.E., Utter, G., Houghten, R. A., and Lerner, R. A. (1989). *J. Immunol.* **143**, 3602–3608.

Närvänen, A., Korkolainen, M., Suni, J., Korpela, J., Kontio, S., Partanen, P., Vaheri, A., and Huhtala, M–L. (1988). *J. Med. Virol.* **26**, 111–118.

Närvänen, A. (1990). *Synthetic Peptides as Probes for Protein Interactions and as Antigenic Epitopes.* Thesis: In *Biol. Res. Rep. Univ. Jyväskylä* **21**, 1–89.

Närvänen, A., Seppänen, H., Korkolainen, M., Hukki–Immonen, O., Leinonen, M., Huhtala, M.–L., and Saikku, P. (1991). In: *Chlamydial Infections of the Genital and Respiratory Tracts and Allied Conditions* (P.–A. Mürdh and P. Saikku, eds.), Gummerus, Jyväskylä, Finland.

Papsidero, L. D., Sheu, M., and Ruscetti, F. W. (1989). *J. Virol.* **63**, 267.

Pfaff, E., Mussgay, M., Bîhm, H. O., Schultz, G. E., and Schaller, H. (1982). *EMBO J.* **1**, 869–874.

Ralston, S., Hoeprich, P. D., and Akita, R. (1989). *J. Biol. Chem.* **264**, 16343.

Roivainen, M., Närvänen, A., Korkolainen, M, Huhtala, M–L., and Hovi, T. (1991). *Virology* **180**, 99–107.

Rosen, J., Hom, Y–L., Whalley, A., Smith, R., and Naso, R. B. (1987). In: *Vaccines*. Cold Spring Harbor Laboratory. Cold Spring Harbor, New York. **87**,188–193.

Schaaper, W. M. M., Lankhof, H. and Meloen, R. H. (1990). *Proc. 21st Europ. Peptide Symp.*, Barcelona.

Schoofs, P. G., Geysen, H. M., Jackson, D. C., Brown, L. E., Tang, X–L., and White, D. O. (1988). *J. Immunol.* **140**, 611.

Shukla, D. D., Tribbick, G., Mason, T. J., Hewish, D. R., Geysen, H. M., and Ward, C. W. (1989). *Proc. Natl. Acad. Sci. USA* **86**, 8192–8196.

Singh, K., Sheridan, P., Aldwin, L., and Nitecki, D. E. (1989). *Proc. 11th Am. Peptide Symp.*, 898.

Stemler, M., Weimer, T., Tu Z. X., Wan, D. F., Levrero, M., Jung, C., Pape, G. R., and Will, H. (1990). *J. Virol.* **64**, 6, 2802–2809.

Tainer, J. A., Getzoff, E. D., Alexander, H., Houghten, R. A., Olson, A.J., and Lerner, R.A. (1984). *Nature* **312**, 127–134.

Thornton, J. M., Edwards, M. S., Taylor, W. R. & Barlow, D. J. (1986). *EMBO J.* **5**, 409–413.

Varnier, O. E, Närvänen, A., Korkolainen, M., Lillo, F., Kontio, S., Elm, J., Suni, J., Vaheri, A., and Huhtala M.–L. (1991). *Ann. New York Acad. Sci.* **626**, 502–515.

Westhof, E., Altschuh, D., Moras, D., Bloomer, A. C., Mondragon, A., Klug, A., and Van Regelmortel, M. H. V. (1984). *Nature* **331**, 123–126.
Wang, J. J. G., Steel, S., Wisniewolski, R., and Wang, C. Y. (1986). *Proc. Natl. Acad. Sci. USA* **83**, 6159–6163.
van der Zee, R., van Eden, W., Meloen, R. H., Noordzij, A., and van Embden, J. D. A. (1989). *Eur. J. Immunol.* **19**, 43.

6

Fast Multiplex Polymerase Chain Reaction on Boiled Clinical Samples for Rapid Diagnosis of Viral Infections

Christian Vandenvelde and Danièle Van Beers

I. INTRODUCTION

In spite of the recent dramatic development of new, rapid methods for detection of viral infections, most laboratory diagnosis of viral disease still depends either on the growth of virus or on the detection of a specific serologic reaction to infection. As tissue culture is the most widely sensitive and flexible, it remains the golden standard. But, because no single monolayer type is sensitive to the growth of all viruses most laboratories choose an array of different tissue cultures and, depending on the suspected viruses, inoculate each specimen into those lines or strains most likely to give a positive result. For the most part each virus, or virus group, produces a characteristic cytopathic effect, and the skilled technologist can usually develop a tentative identification on the basis of cell tropism and morphology, giving the clinician useful information at the earliest possible moment. Final identification of virus isolates usually depends on serologic tests and sometimes on morphologic, biophysical, or biologic characterization. The polymerase chain reaction (PCR) has the potential to replace these laborious viral isolation methods, but problems such as specificity, reproducibility, rapidity, cost, and the need for a common sequence in all strains of a particular virus must be taken into account. The production of low and high molecular weight non–specific PCR products must be noted while using most published PCR protocols. The accumulation of such PCR–products depletes primers and deoxynucleotide–triphosphates from the reaction mixture and leads to competition for enzyme with the desired PCR–product. Diagnosis must then rely on final detection assays such as hybridization with radioactive oligonucleotide probes to guarantee sufficient specificity, sensitivity and reproducibility.

Christian Vandenvelde and Danièle Van Beers • Hôpital Universitaire Brugman, Université Libre de Bruxelles, Bruxelles, Belgium.

Applied Virology Research, Vol. 3, Edited by
E. Kurstak *et al.*, Plenum Press, New York, 1994

A highly efficient PCR procedure was devised for rapid viral detection in clinical samples that is not dependent on complicated recombinant DNA methodology.

In doing so, the primary goal was to complement routine clinical screening. As, unlike cell culture, PCR will only detect the virus for which the primers are designed, the new procedure was adapted to the simultaneous amplification of four different DNA sequences in a single rapid reaction. In this chapter, the optimal reaction conditions for making the PCR technique as efficient and practical as possible for routine applications are described, using the simultaneous detection of DNA sequences from the three most prevalent high risk genital papillomaviruses and the detection of hepatitis B virus at its infectivity threshold.

Scientists from all over the world have discussed the link between these viruses and human cancer for 15 years. For anogenital cancers, the consensus view favours an aetiologic role for some HPVs. This opinion is supported by epidemiological studies showing that specific virus types, such as HPV 16 and 18, are present in many squamous cancers of the genital tract, and by a large body of experimental evidence showing that HPV 16, unlike HPV 6, has transforming properties and can immortalise primary human keratinocytes *in vitro*. However, there is general agreement that the relation between virus and tumour is not as straightforward as it had seemed when the association was first reported 15 years ago. The main reason for reservations about the role of papillomaviruses is the conflicting epidemiological studies that are said to indicate the oncogenicity of HPV 16. In a review of the evidence, a leading epidemiologist drew attention to the extraordinary range of prevalences assigned to HPV 16 by twenty different laboratories in various part of the world (Munoz, 1988). The reported HPV 16 detection rates in cervical cancer biopsies in the UK, Germany, the USA, and Japan ranged from 18 to 100 %, while the HPV 16 detection rate in normal cervix ranged from 0 to 80 %. These results might indicate either that the prevalence of HPV 16 varies in different parts of the world, or that there must exist important differences in study design (selection bias, sample size, choice of controls) and in sensitivity and specificity of the hybridisation techniques used (Southern blot, dot blot, *in situ* filter hybridisation, PCR). The influence of DNA technology on HPV prevalence surveys has been illustrated recently by two groups in the UK who used PCR to detect the virus. In a series of supposedly controlled experiments designed to exclude the possibility of sample–to–sample contamination and PCR–product carryover, Young *et al.* (1989) and Tidy *et al.* (1989) showed that HPV 16 DNA is present in 100 % of cancer biopsies and in 80 % of normal cervical scrapes, indicating that the rate of infection with HPV 16 in the general population is very high. Preliminary reports from centres in the USA confirmed the specificity and sensitivity of their approach (Manos *et al.*, 1989; Beckmann *et al.*, 1989). These findings highlight the fact that classical PCR protocols can not be used for epidemiological studies and routine viral diagnosis. By comparison, the HPV 16, 18 and 33 detection rates found when performing the Fast Multiplex PCR procedure on normal, dysplastic and neoplastic cervical scrapes, were situated between those reported when using *in situ* filter hybridization and those reported when using classical PCR protocols, indicating that Fast Multiplex PCR is both more sensitive and more specific that existing DNA detection assays.

Since 10 to 15% of post–transfusion viral hepatitis cases are caused by hepatitis B virus, and despite mandatory third generation screening procedures for HBsAg, there is an urgent need for a simple, time–cost–effective, but very sensitive test for routine HBV DNA detection in serum. Commercially available (Genostics™ Hepatitis B Viral DNA assay, Abbott Laboratories) and laboratory designed (Scotto *et al.*, 1983) hybridization assays are not sensitive enough (10^5 and 3×10^4 virus particles/ml, respectively). The sensitivity of PCR–Southern blotting assays on purified HBV DNA ranges from 10 (Ulrich *et al.*, 1989) to 3×10^2 (Larzul *et al.*, 1988) virus particles/ml serum. A two–step PCR assay of purified HBV DNA (Larzul *et al.*, 1990) has been shown to detect HBV at the infectivity threshold in serum (10^2 virus particles/ml). DNA purification is, however, too labour–intensive for routine testing. Unfortunately, PCR–Southern blotting assay and nested–primer PCR (Kaneko *et al.*, 1989) on either proteinase K or NaOH pretreated serum could only detect HBV virions when above the 10^3 and 3×10^2 levels, respectively. Serum protein had a very strong inhibitory effect on *Thermus aquaticus* (Taq) DNA polymerase. Preliminary experiments using the Fast PCR procedure on serum diluted ten–fold in PBS, pH 7.2, has demonstrated that PCR could achieve reproducible virus detection toward the 2×10^3 virus particles/ml level. Thirty–five serum pretreatments have been attempted with the aim of reaching a reproducible detection limit at the infectivity threshold of HBV in serum. This target level was easily reached using sodium octanoate thermoprotection as pretreatment and Fast PCR for 99 cycles, while the classical phenol/chloroform/alcohol DNA purification procedure reached the 10 virus particles/ml level. Because of its greater simplicity and its lower risk of sample–to–sample cross contamination than either DNA purification or NaOH pretreatment, we recommend the use of the sodium octanoate thermoprotection method for routine clinical detection of HBV in serum.

II. THEORETICAL AND EXPERIMENTAL BACKGROUND

As a starting point for optimization, the recommendations given in the booklet of the Gene Amp DNA Amplification Reagent kit from Perkin–Elmer Cetus (Norwalk, USA) were followed. After an initial thermal denaturation of the purified DNA template at 94°C for 1.5 minutes, the following cycle profile was suggested: 2 minutes at 37°C for primer annealing, 3 minutes at 72°C for amplimer extension, and 1 minute at 94°C for DNA denaturation. At the end of the 25th cycle, the heat denaturation step was omitted and the extension step was lengthened by an additional 7 minutes. The sensitivity and specificity of this protocol was evaluated by assaying various primers–template systems in a wide range of reaction conditions (Vandenvelde *et al.*, 1990). All possible parameters were modified except for the thermo–cycling segment times. Using the starting procedure, a specific segment of DNA could be amplified by as much as 10^5 fold, but there was evidence of non–specificity as seen by the presence of background DNA of higher molecular weight than the target sequence. As expected, increasing the initial denaturation step to 15 minutes and the number of cycles to 30 resulted in greater amplification (10^7 fold), but non–specific bands still occurred. At an annealing temperature of 55°C, significantly better specificity was achieved without loss of sensitivity. However, neither specific nor non–specific bands were obtained when using some pairs of primers, and it was also noted that some target bands became fainter. The production of low and high molecular weight products was still frequently observed while using the optimized protocol. This was most obvious after many cycles, in the presence of high primer, deoxynucleotide or Taq DNA polymerase concentrations, at low starting template DNA concentration, with lowered annealing temperature and prolongated annealing time. Taq DNA polymerase, even in the absence of any primer complementarity, may transiently hold two single–stranded DNA molecules, such that their 3' ends are adjacent and may be extended. Accumulation of such PCR–products depletes primers and deoxynucleotides from the reaction mixture and leads to competition for enzyme with the desired PCR–product. Thus, decreasing primer, deoxynucleotide and enzyme concentrations, increasing annealing temperature and reducing annealing incubation time would suppress the primer–dimers production during the annealing steps. Because of the higher annealing temperature and the shorter annealing incubation time, the formation of perfect hybrids between the extension primers and the template will be favoured. Moreover, the higher temperature will reduce the extent of secondary structures in the template, thereby potentially minimizing spurious polymerisation. With respect to reproducibility, four factors appeared to be very important: the sample pretreatment, the denaturation temperature, the amplimer length, and the use of a recombinant Taq DNA polymerase. For all primer–template systems tested, the best results were always obtained after prolonged thermal destruction of the sample (minimum 99°C for minimum 15 minutes). By contrast, DNA extraction, proteinase K and NaOH pretreatment, did not permit reproducible results. Complete strand separation was always achieved above 93°C, whatever the denaturation segment time programmed. One may thus reduce the denaturation time, but must be sure that an adequate temperature has been reached. Adding an oil overlay to the reaction mixture reduces both cooling and liquid loss due to evaporation, particularly at the denaturation step. Indeed, samples without oil do not reach the denaturation temperature, presumably because of evaporative cooling. Too much oil slows the rate of approach. Two drops of light mineral oil (Sigma Chemical Co.) can control the concentration of the components and the uncertainty in fluid temperature, leading to improved PCR yields and reproducibility. A denaturation step at 95°C seems to be optimal because there may be some reduction in enzyme activity above this temperature. Slightly higher PCR–product yield, specificity and reproducibility of amplification were obtained with recombinant than with native Taq DNA polymerase. A purer

DNA polymerase is indeed ensured by recombinant technology because contamination by nucleases or other thermostable enzymes do not occur.

Innis *et al.* (1988) have reported an extension rate of 60 nucleotides/second/enzyme molecule at 70°C and significant extension activity at 55°C (24 nucleotide/second/enzyme molecule) with Taq DNA polymerase. At a temperature higher than 85°C, DNA synthesis may be limited by the stability of the primer–template duplexes. When using the DNA Thermal Cycler from Perkin–Elmer Cetus, the maximal programmable rate for thermal ramps is one degree per second. So, if one omits the extension plateau segment at 72°C, the transition time between an annealing temperature set at 55°C and a denaturation temperature set at 95°C will be exactly 40 seconds, and efficient primer extension will occur during the first 30 seconds of this transition segment. It proved sufficient to amplify target sequences up to 500 base pairs long. But this was only true when the extension primers were correctly chosen. A good amplification primer should form stable duplexes with the target sequence under the annealing conditions, be highly specific for the intended target sequence (not base–pairing to other regions within the template or to other templates), and not anneal to itself. The critical component in the search for oligonucleotides which would optimally meet all three of these criteria is the algorithm used for determination of the duplex dissociation or melting temperature (Tm). An algorithm which is commonly used is that of Suggs *et al.* (1981) in which Tm calculation is based on the number of AT and GC base pairs. As base sequence not base composition determines stability, a more precise method for determination of Tm is based on nearest–neighbor thermodynamic parameters (Breslauer *et al.*, 1986).

The melting behaviour of any DNA duplex structure can indeed be predicted from its primary sequence if the relative stability (free energy) and the temperature–dependent behaviour (enthalpy, entropy) of each DNA nearest–neighbor interaction are known. This predictive ability proved valuable for calculating the minimum length of a primer required to form a stable duplex with a target gene at a given annealing temperature (20–mer for annealing at 55°C), for estimating the optimum melting temperature of a duplex structure formed between a 20–mer primer with random base distribution and its complementary gene segment (85.5 ± 12.5 °C), and for identifying potential sites of local melting within a polymer duplex by predicting the sequence–dependent melting temperatures of local DNA domains (GGG, CCC, TATA, ATAT, TTTT, CTTG and TTTA should always be avoided, particularly at the 3' end of the primer). If primers containing these low melting–temperature DNA domains have to be used for some diagnostic applications, their concentrations must be adjusted. Indeed, when the same number of copies of three HPV target sequences were run simultaneously in a single tube in the presence of three pairs of HPV type specific primers, specific amplifications occurred preferentially for some primers target systems. This phenomenon was reproducible for any deoxynucleotide and $MgCl_2$ concentrations and for whatever annealing temperature was chosen. It did not correlate with amplimer or primer length, nucleotide content or melting temperature. On the other hand, it correlated well with the number of 5'–AT–3' and 5'–TA–3' sequences contained in the primers. Using a two–fold decreased concentration of the primers with the lowest 5'–AT–3' and 5'–TA–3' sequence content and a two–fold increase concentration of the primers with the highest 5'–AT–3' and 5'–TA–3' sequences content, reproducible simultaneous amplification of three or four target sequences was achieved without modifying the magnesium concentration. Thus, preference should be given to a random base sequence distribution with an average GC–content and a low 5'–AT–3' and 5'–TA–3' sequences content, the presence of guanosine or cytosine at the 3' (and 5') end, the absence of primer complementarity and secondary structure, and a calculated melting temperature of about 60°C by Suggs algorithm (Suggs *et al.*, 1981) and 85°C by Freier–Breslauer algorithm (Freier *et al.*, 1986).

The primers must also be selected for minimal homologies with the foreign DNA sequences detailed in the Gene Bank database by the Goad and Kanehisa algorithm (Goad and Kanehisa, 1982). Finally, the primers should be tested for specificity in the presence of DNA sampled from several individuals, bacteria, fungi and viruses. A typical PCR reaction can generate 10^{12} molecules of amplified DNA that can potentially contaminate subsequent amplifications of the same target sequence. This type of contamination has been termed PCR–product carryover to differentiate it from contamination by sample DNA. Using the same principles applied to sterile handling of RNA or cell cultures (which also expand exponentially) PCR–product carryover or sample cross–contamination was avoided. Indeed, positive results were never obtained from negative control samples as a result of contamination with positive template DNA. All the precautions to be taken in setting up the PCR reactions are described elsewhere by Kwok and Higushi (1989). Because cross–contamination between samples can also be a factor, the following additional precautions should be taken: change gloves as frequently as possible, minimize sample handling, select a sample that amplifies weakly but consistently for use as a positive control, and above all avoid using plasmid DNA. Ten to one hundred cells from cell lines, known to be positive or negative for the concerned sequences, can be easily used as controls.

All the data above strongly suggest that one might reduce annealing and denaturation hold times to one second per cycle and completely suppress extension hold times without anxiety for loss of PCR efficiency, because primer annealing is achieved in 10 milliseconds (Saiki, 1989), because sufficient time is given for primer extension, and because denaturation time is less important than the temperature reached at the end of each cycle. To the contrary, this "turbo" thermo–cycling profile has been shown to be at least 2,000–fold more sensitive than classical PCR–protocols, 100% specific and 100% reproducible in several microbiological and haematological routine applications such as HPV types 16, 18 and 33 and *Chlamydia trachomatis–Neisseria gonorrhoeae* cervical screening (Vandenvelde *et al.*, 1990), hepatitis B virus detection at the infectivity threshold in serum (Vandenvelde *et al.*, 1993), parvovirus B19 detection in serum, *Toxoplasma gondii* detection in amniotic and cerebrospinal fluids (data to be published), detection of minimal residual disease in B–cell malignancies by amplification of the monoclonal rearrangements of the immunoglobulin heavy

Table 1. Human Papillomavirus Type–Specific Amplification Primers

Name Concentration	Sequence 5′ – 3′	Mer bp	Tm °C	AT/TA %	Band bp	Target Sequence
PCR161 60 pM	CTAAGGCCAACTAAATGTCAC	21	60	15	221	HPV16–LCR Screening test
PCR162 60 pM	GTCTGCTTTATACTAACCGG	20	58	21		
PE7181 30 pM	ATTCCGGTTGACCTTCTATG	20	58	11	197	HPV18–E7ORF Screening test
PE7182 30 pM	GAATGCTCGAAGGTCGTCTG	29	62	5		
PE7331 30 pM	GGCTTGGACCGGCCAGATGG	20	68	5	178	HPV33–E7ORF Screening test
PE7332 30 pM	GTGCACAGGTAGGGCACAC	19	62	6		
PE7161 30 pM	GCAGAACCGGACAGAGCC	18	60	0	127	HPV16–E7ORF Confirmatory test
PE7162 30 pM	GTGTGCCCATTAACAGGTC	19	58	11		
PE7183 30 pM	GCCCGACGAGCCGAACCAC	19	66	0	109	HPV18–E7ORF Confirmatory test
PE7184 30 pM	GGAATGCTGAAGGTCGTC	19	60	5		
PE7333 60 pM	AGGATGAAGGCTTGGACCG	19	60	6	156	HPV33–E7ORF Confirmatory test
PE7334	GTGCCCATAAGTAGTTGCTG	20	60	16		
PC03 30 pM	ACACAACTGTGTTCACTAGC	20	58	5	110	Human β-globin gene (first exon) Screening test
PC04 30 pM	CAACTTCATCCACGTTCACC	20	60	5		

chain gene (Vandenvelde *et al.*, 1991a), and in T–cell malignancies by amplification of the monoclonal rearrangements of the T–cell receptor delta and gamma chains genes (Vandenvelde and Van Beers, 1991b). Moreover, the average cycle efficiency was estimated at 75% for 40 cycles of Fast–PCR and at 50% for the 40 cycles optimized Gene Amp PCR protocol. A Fast PCR cycle is however achieved in 80 seconds, while a Gene Amp PCR cycle requires 7 minutes 54 seconds for completion. These data strongly suggest that all the statements made above concerning annealing, extension and denaturation hold times are correct. The whole procedure requires less than 4 h to be performed from which only about 1 h for PCR process, indicating that it is not only the most efficient but also the simplest and quickest way to carry out the PCR technique.

Indeed, neither the hot start PCR procedures, nor the presence of uracil–N–glycosylase and dUTP in the PCR reaction mix, nor the nested and double–nested primer PCR procedures, nor the use of radioactive, enzyme–labeled and chemiluminescent probes after PCR produced better results in routine applications. A reciprocal correlation between the concentration of hydrolyzed proteins in the amplification reaction and amplification efficiency has been observed either with serum or with 7% bovine serum albumin (Ulrich *et al.*, 1989). Phenol–chloroform extraction of digested proteins followed by dialysis and concentration of the aqueous phase in a microconcentrator was thus considered as the procedure of choice for the recovery of low amounts of HBV DNA from serum (Ulrich *et al.*, 1989). From the 30 published simplified procedures we have assessed, the NaOH extraction method (Kaneko *et al.*, 1989) was found the most sensitive (100 virus particles/ml) but insufficiently reproducible and specific (Vandenvelde *et al.*, 1993). It is indeed difficult to warrant complete neutralization of NaOH without pH measurement and absence of sample–to–sample cross contamination when first adding pretreated serum to the reaction tube. This very simple and sensitive method had thus to be considered as unreliable for routine clinical detection of HBV infection. Industrial albumin purification from plasma by thermocoagulation is based almost exclusively on the fact that, unlike other plasma proteins, albumin can be stabilized against thermal denaturation by low concentrations of short chain fatty acids, in particular octanoic (caprylic) acid (Hoch and Chanutin, 1954). In addition to being a stabilizer for albumin, caprylate acts as a precipitating agent for all denatured proteins. Because in our experience bovine serum albumin had an inhibitory effect on PCR amplification while gelatin improved PCR–product yield, we suspected denatured albumin to be responsible for most of the inhibitory effect of serum on Taq DNA polymerase. This was verified by stabilizing serum albumin with 0.08 M caprylic acid before sample thermal denaturation at 99°C for 15 min. The results suggested that the hypothesis was correct — using this simple technique, we were able to routinely detect one HBV genome in the reaction mixture.

III. FAST MULTIPLEX PCR–BASED SCREENING FOR HIGH RISK GENITAL PAPILLOMAVIRUSES

Normal (323) and dyskaryotic (71) samples from women attending either the cervical cancer screening clinic (336) or the routine colposcopy clinic (58) of Brugmann University Hospital in Brussels were included in this screening study. All women agreed to participate in the study. Patients were classified as normal (CINO) or dyskaryotic (CIN 1, CIN 2, CIN 3/CIS) on the basis of exfoliative cytology and scrapes for DNA analysis were taken immediately following scrapes for cytological examination, using plastic Eyre's spatules. Wood spatules are not recommended because wood seems to contain inhibitory substances for Taq DNA polymerase. Spatules for HPV screening were placed in separate disposable plastic bags, where epithelial cells were washed away using 1 ml PBS pH 7.2. Aliquots (10 μl) containing approximately 500 cells/μl were prepared and assayed immediately or kept at –20°C. Cervical punch biopsies were performed to confirm viral and dysplastic changes found at cytologic examination.

Well characterized cell lines such as MRC–5, Caski (Baker *et al.*, 1987), and HeLa (Popescu *et al.*, 1987) that contain no HPV, HPV 16 or HPV 18 DNA sequences were cultured using classical techniques. *E. coli* colonies containing HPV 33 genome inserted in a pBR 322 plasmid were cloned using well–known techniques. Aliquots containing either 100 MRC–5 or 10 Caski, 100 HeLa and 5,000 pBR 322–HPV 33 in 10 μl PBS pH 7.2 were prepared, kept at –20°C, and used as internal and quadruple positive controls, respectively. HPV 33–plasmids were replaced as soon as possible by cells from a cervical sample found positive with two different sets of HPV 33 specific primers.

All the primers used are detailed in Table 1 and were chosen to anneal either the E6 ORF, or the E7 ORF, or the long control region (LCR) of the concerned papillomaviruses. Primers for the human β–globin gene were used as internal control (Saiki *et al.*, 1986). Four pairs of primers were used to screen cervical scrapes for high risk genital HPV infection; three other pairs of primers were used to confirm HPV detection.

After an initial DNA template thermal denaturation at 99°C for 15 minutes in a 1.5 ml Eppendorf tube containing a 100 μl light mineral oil overlay, the following cycle profile was programmed on the Perkin Elmer Cetus DNA Thermal Cycler: one second at 55°C (annealing–extension step) and one second at 95°C (denaturation step). An extension step at 72°C for 10 minutes took place at the end of the 40th cycle. All samples were assayed in duplicate.

Fast Multiplex PCR was performed in a 50 μl volume and, in addition to 10 μl boiled sample DNA, contained 50 mM NaCl, 10 mM Tris–HCl (pH 8.3 at 25°C), 3 mM $MgCl_2$, 100 μg/ml gelatin, 30 picomoles of each of the 8 primers, 200 μM of each deoxynucleotide triphosphate, 1 unit of Ampli Taq (recombinant Taq DNA polymerase from Perkin–Elmer Cetus), and 5% (v/v) dimethyl–sulfoxide (DMSO). The components were added in a 0.5 ml Eppendorf tube chilled on ice in the following order: 1) two drops of light mineral oil; 2) 11 μl of 5X PCR buffer

Table 2. High Risk Genital Papillomavirus Prevalence in the Belgian Female Population by CIN Status

%	CIN 0	CIN 1	CIN 1 or 2	CIN 2	CIN 2 or 3	CIN 3/CIS
HPV ±	96.13	1.19	2.38	1.19	2.68	1.49
HPV –	86.69	50.00	42.22	33.33	19.15	7.69
HPV +	13.31	50.00	57.78	66.67	80.85	92.31
HPV 16	4.95	29.17	35.56	42.86	51.06	57.69
HPV 18	1.86	16.67	13.33	9.52	12.77	15.38
HPV 33	7.43	16.67	17.78	19.05	19.15	19.23

containing DMSO and Ampli Taq; 3) 6.4 µl of each of the four primer–mixes; 4) double distilled water to a volume of 40 µl; and 5) 10 µl boiled sample DNA. As explained in the preceding section, the concentrations of the primers were not always set at 30 picomoles per reaction (see Table 1). DMSO has been used to linearize template DNA, because complex secondary structures in the template DNA may hinder the extension of the primers by the polymerase, and because GC–rich DNA may impede initial genomic denaturation, as well as sub–genomic denaturation of the DNA, which is crucial for the efficient annealing of the primers to the template. Reaction yields were evaluated on 3% Nu Sieve GTG (FMC Corporation, Maine, USA) agarose gels stained with ethidium bromide (Maniatis *et al.*, 1982). The gels were interpreted without prior knowledge of the cytological findings.

Preliminary experiments have shown that 1 Caski, 10 HeLa and 500 pBR 322–HPV 33 could be detected simultaneously by Fast Multiplex PCR in the presence of 10^4 MRC–5. Since routine experiments are performed on about 5,000 cells potentially infected cervical cells, the sensitivity of the technique is more than sufficient (amplification factor = 10^9 after 40 cycles). The results of the duplicated detection and confirmation assays agreed in all cases. Positive results were never obtained from negative control samples as a result of a contamination with positive template DNA. The production of low or high molecular weight non–specific PCR–products was never noted. Sometimes, no PCR–product from the positive controls nor from the samples could be seen on the agarose gel. When positive controls failed, the experiment was performed again, but with a new batch of Ampli Taq DNA polymerase. The results were then always satisfactory.

Some patients were examined more than once for reasons varying from routine diagnostic tests to re–examination of existing lesions. These patients were classified according to the most abnormal cytological finding, and as positive for HPV infection if the Fast Multiplex PCR results were positive on this occasion. The percentage of cervical scrapes found negative for the three HPVs (HPV–), the percentage of scrapes found positive for one or more of the three HPV s, and the HPV 16, HPV 18 and HPV 33 detection rates according to cytological findings (CINO, CIN1, CIN1 or 2, CIN2, CIN2 or 3, CIN3/CIS) are shown in Table 2. The prevalences of the different degrees of dysplastic changes in the cervix of the women attending the cervical cancer screening clinic (HPV +/–) are also given in Table 2. The detailed statistical anlysis of these data in the light of other well–known risk factors have been partially published (Vandenvelde *et al.*, 1992; Vandenvelde and Van Beers, 1992, 1993). Since all positive results were reproducible and could be confirmed using a second pair of HPV type specific primers, and since the technique was sufficiently sensitive to allow the detection of 100 to 500 viral copies in the presence of 10^4 cells, false positives and false negatives are highly improbable. So, the prevalence data presented here are probably the most accurate to date. The HPV 16, 18 and 33 detection rates found here are indeed situated between those reported when using *in situ* filter hybridisation and those reported when using classical PCR protocols, suggesting that Fast Multiplex is both more sensitive and more specific than all existing DNA detection assays.

IV. CONCLUSION

In all applications tried to date in our hospital–situated laboratory, the Fast (Multiplex) PCR thermo–cycling method remained the most sensitive, specific, reproducible and time–cost–effective for routine clinical investigation.

REFERENCES

Baker, C. C., Phelps, W. C, Lindgren, V,. Braun, M.J., Gonda, M.A. and Howley, P.M. (1987). *J. Virol.* **61**, 962–971.

Beckmann, A.M., Chu, J., Miller, B.A., Christiansen, A.E., McDougall, J.K. and Galloway, D.A. (1989). *Proc. Eighth Intl. Papillomavirus Workshop.* Taos, New Mexico.

Breslauer, K.V., Frank, R., Blöcker, H. and Marky, L. A. (1986). *Proc. Natl. Acad. Sci. USA.* **83**, 3746–3750.

Freier, M.S., Kierzek, R., Jaeger, J.A., Sugimoto, N., Caruthers, M. H., Neilson, T. and Turner, D. H. (1986). *Proc. Natl. Acad. Sci. USA.* **83**, 9373–9377.

Goad, W. B. and Kanehisa, M. I. (1982). *Nucleic Acids Res.* **10**, 265–278.

Hoch, H. and Chanutin, A. (1954). *Arch. Biochem. Biophys.* **51**, 271–276.

Innis, M. A., Myambo, K. B., Gelfand, D. H. and Brow, M. A. D. (1988). *Proc. Natl. Acad. Sci. USA.* **85**, 9436–9440.

Kaneko, S., Feinstone, S. M. and Miller, R. (1989). *J. Clin. Microbiol.* **27**, 1930–1933.

Kwok, S. and Higuchi, R. (1989). *Nature* **339**, 237–238.

Larzul, D., Guigne, F., Sninsky, J., Mack, D., Bréchot, C. and Guesdon, J. (1988). *J. Virol. Method.* **20**, 227–237.

Larzul, D., Chevrier, D., Thiers, V. and Guesdon, J. (1990). *J. Virol Method.* **27**, 49–60.

Maniatis, T., Fritsch, E. F. and Sambrook, J. (1982). *Molecular Cloning: A Laboratory Manual.* Cold Spring Harbor Laboratory Press, Cold Spring Harbor.

Manos, M. M., Wright, D. K., Lewis, A. J., Wolinsky, S., Broker, T. R. and Ting, Y. (1989). *Proc. Eighth Intl Papillomavirus Workshop,*.Taos, New Mexico.

Munoz, N., Bosch, X. and Keldor, J. M. (1988). *Br. J. Cancer* **57**, 1–5.

Popescu, N. C., Di Paolo, J. A. and Amsbaugh, S. C. (1987). *Cytogenet. Cell Genet.* **44**, 58–62.

Saiki, R. K., Bugawan, T. L., Horn, G. T., Mullis, K. B. and Erlich, H. A. (1986). *Nature* **324**, 163–166.

Saiki, R. K. (1989). In *PCR technology: principles and applications for DNA amplification.* (H. A. Erlich, ed.). pp. 1–14. Stockton Press, London.

Scotto, J., Hadchouel, M., Hery, C., Yvart, J., Tiollais, P. and Bréchot, C. (1983). *Hepatology* **3**, 279–284.

Suggs, S. V., Hirose, T., Migaken, E. H., Kawashima, M. J., Johnson, K. I. and Wallace, R. B. (1981). In *ICN–UCLA Symp Dev Biol Using Purified Genes.* (D. D. Brown, ed.). pp. 683–693. Academic Press Inc, New–York.

Tidy, J. A., Parry, G. C. N. and Ward, P. (1989). *Lancet* **i**, 434.

Ulrich, P., Bath, R., Seto, B., Mack, D., Sninsky, J. and Vyas, G. (1989). *J. Infect. Dis.* **160,** 37–43.

Vandenvelde, C., Scheen, R., Corazza, F. and Van Beers, D. (1991a). *J. Exptl. Clin. Hematol.* **33,** 293–297.

Vandenvelde, C., Scheen, R., Defoor, M., Duys, M., Dumon, J. and Van Beers, D. (1993). *J. Virol. Method.* **42,** 251–264.

Vandenvelde, C., Scheen, R., Fondu, P. and Van Beers, D. (1991b). *J. Exptl. Clin. Hematol.* **33,** 25–29.

Vandenvelde, C., Scheen, R., Van Pachterbeke, C., Loriaux, C., Decelle, J., Hubert, T., Delhaye, C., Cattoor, J–P., Duys, M. and Van Beers, D. (1992). *J. Med. Virol.* **36,** 279–282.

Vandenvelde, C. and Van Beers, D. (1992). *J. Med. Virol.* **38,** 226–232.

Vandenvelde, C. and Van Beers, D. (1993). *J. Med. Virol.* **39,** 273–277.

Vandenvelde, C., Verstraete, M. and Van Beers, D. (1990). *J. Virol. Methods* **30,** 215–227.

Young, L. S., Bevan, J. S. and Johnson, M. A. (1989). *Br. Med. J.* **298,** 14–18.

7

Digoxigenin Labeled Probes and their Use in the Laboratory Diagnosis of Virus Infections

Pirkko Heino, Veijo Hukkanen, and Pertti P. Arstila

I. INTRODUCTION

The last few years have seen a rapid spread of a novel nonisotopic label, digoxigenin (DIG), in use for nucleic acid hybridizations. Nucleic acid hybridization has been, and still is, the basic method in modern molecular biology. Most applications involve the use of radiolabeled–probes. Radioisotope labels are sensitive and well established for laboratory use, but on the other hand the labels are short–lived and special facilities and equipment are required for handling of isotopes. In most countries there are strict regulations for the use of radioactive materials.

Isotopic labels have been partly replaced by biotin–avidin/streptavidin labeling systems (Langer *et al.*, 1981). In most cases, the sensitivity is not as high as with isotopic labels, and the user of biotin may encounter the problem of background staining, due to the presence of endogenous biotin in some tissues (Herrington *et al.*, 1989a, Morey *et al.*, 1992). The digoxigenin–labeling method was originally developed by the Boehringer Mannheim Corporation, (Martin *et al.*, 1990). DIG–probes clearly overcome many problems associated with other labels.

II. PRINCIPLE OF THE DIGOXIGENIN LABELING SYSTEM

Digoxigenin (DIG) is a steroid hapten, a derivative of the cardiac glycoside digoxin, which is conjugated to deoxyuridine triphosphate (dUTP) by an 11

Pirkko Heino • Department of Virology, Karolinska Institute, Stockholm, Sweden. *Veijo Hukkanen and Pertti P. Arstila* • Department of Virology, University of Turku, Turku, Finland

Applied Virology Research, Vol. 3, Edited by
E. Kurstak *et al.*, Plenum Press, New York, 1994

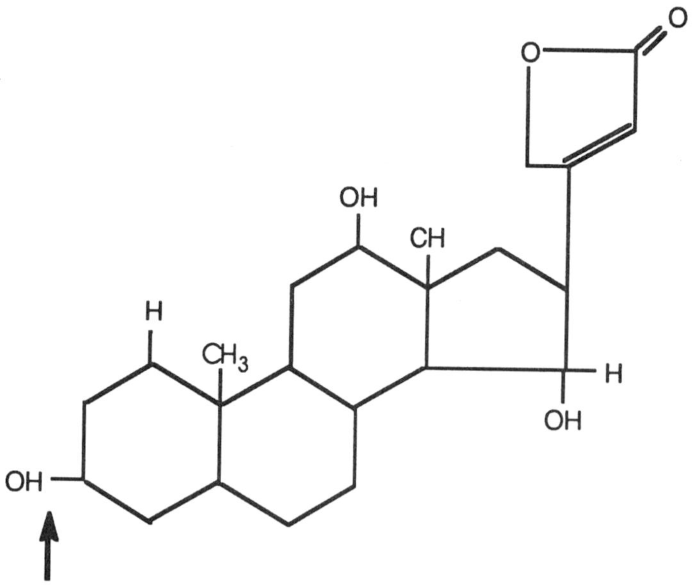

Figure 1.
The structure of digoxigenin. A linker arm is conjugated at the site indicated (arrow).

carbon linker arm (Fig. 1). Digoxigenin occurs naturally in digitalis plants (*Digitalis purpurea* and *Digitalis lanata*) (Lion and Haas, 1990). A great many different DIG–labeling methods have been introduced and are now commercially availabl e for nucleic acid probes, including the random–primed method, nick translation, oligonucleotide tailing, photodigoxigenin, cDNA synthesis (Martin *et al.*, 1990), synthesis of single–stranded DNA probes (Stürzl *et al.*, 1992), incorporation of DIG–labeled nucleotide into polymerase chain reaction (PCR) products (Chikhaoui *et al.*, 1992; Yang *et al.*, 1993; Zakrzewska *et al.*, 1993; Nuovo *et al.*, 1993) and synthesis of DIG–labeled RNA probes (Hukkanen *et al.*, 1990; Heino and Hukkanen, 1992). Labeled–probes are used for Southern, Northern and dot blots, in library screening, for detection of PCR products, and especially in *in situ* hybridization. The detection system is typically based on the use of anti–DIG antibody, which in turn is conjugated to alkaline phosphatase, peroxidase, fluorescein or rhodamine. A chemiluminescent substrate for alkaline phosphatase (Lumigen PPD, Boehringer Mannheim) is also available. The widely used alkaline phosphatase label with the colorimetric substance NBT produces a dark purplish brown precipitate at the location of bound DIG–labeled probe (Martin *et al.*, 1990). A representative staining pattern in nuclei, typical for a given state of human papillomavirus (HPV) infection is illustrated in Fig. 2.

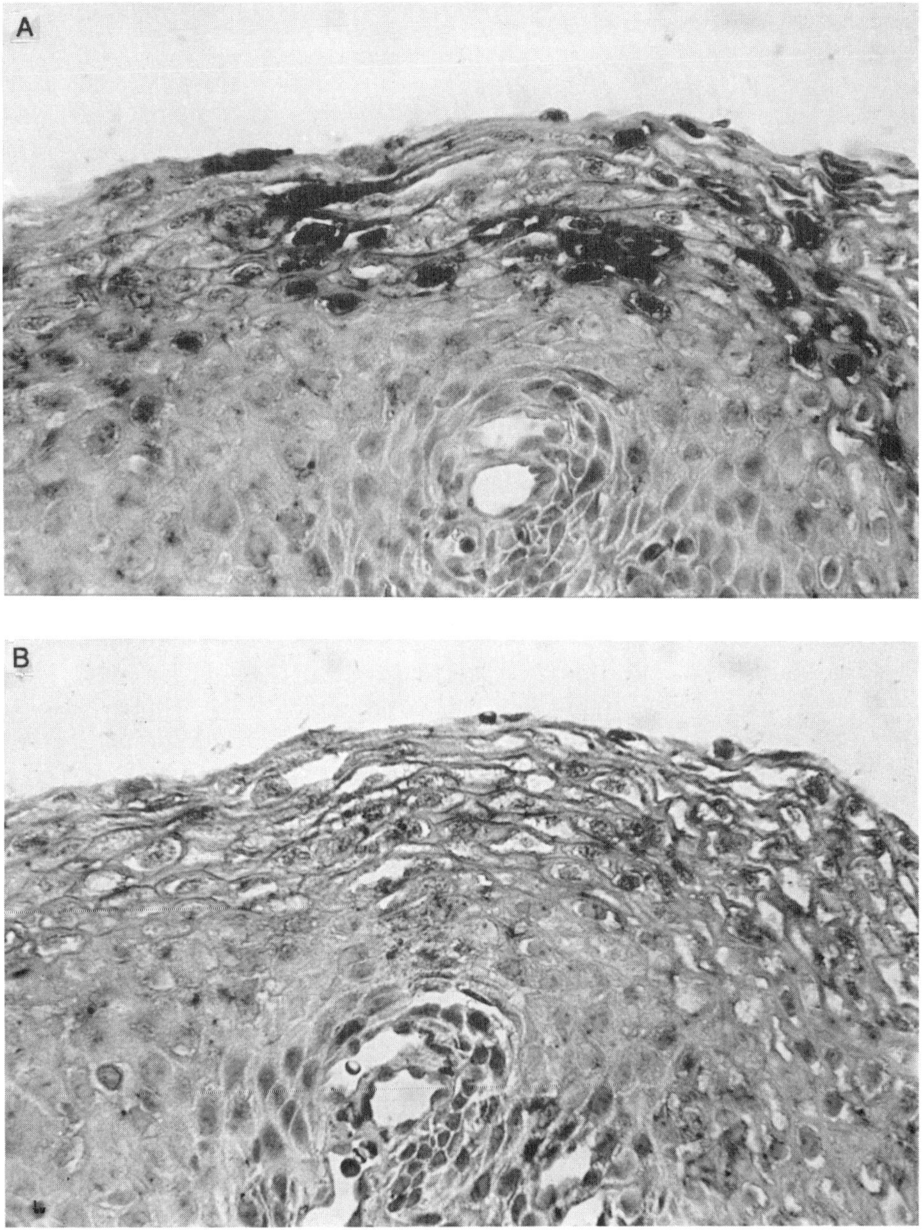

Figure 2.
In situ hybridization of a biopsy specimen from cervical condyloma acuminatum, positive for HPV 11. Hybridization under stringent conditions with digoxigenin-labeled HPV 11 DNA probe (A) and digoxigenin-labeled HPV 16 DNA probe as negative control (B).

III. APPLICATIONS OF DIGOXIGENIN–LABELING IN EXPERIMENTAL AND CLINICAL VIROLOGY

A. In situ hybridization (ISH)

The localization of nucleic acids in tissues by *in situ* hybridization is an essential application for non–radioactive probes. These enable a better resolution at the cellular level in comparison with isotopic labels which emit a range of radiation, thus producing signals also outside the exact location of the probe. Nonisotopic labels can shorten the duration of signal detection, which requires an exposure of days to months using radiolabeled probes.

The first report on the use of a DIG–label in an ISH application for viral nucleic acids was published in 1958 (Heiles *et al.*, 1988), in a study of human papilloma virus (HPV) type 18 in HeLa and HPV 16 in SiHa cell lines. The sensitivity of the digoxigenin label was found to be equal to biotin in a study of HPV DNA in CaSki cells, which contain integrated HPV 16 genomes (Herrington *et al.*, 1989a). The same group also applied the ISH method for simultaneous detection of DIG–labeled HPV 6 and biotinylated Y chromosome in peripheral lymphocytes and human tissues (Herrington *et al.*, 1989b). The DIG–ISH method was first applied for detection of HPV 11, 16 and 18 DNA in genital biopsy specimens by Heino *et al.* (1989). In comparison with radiolabeled probes, the sensitivity of the DIG–DNA probe was equal and the background staining was very low. Using digoxigenin, the localization of hybrids at the cellular level was also better (Heino *et al.*, 1989; Heino and Hukkanen, 1992). Other reports on the use of DIG–DNA probes in ISH for HPV studies include those by Konno *et al.* (1990), Furuta *et al.* (1990), and Morris *et al.* (1990), who found the method to be 4–fold more sensitive in detecting HPV 16 DNA in anal carcinomas than biotin–labeled probes. The excellent localization of nucleic acids in cells by DIG–probes has enabled classification of different types of intranuclear HPV–signals in biopsy specimens, which may have importance in assessment of the state of HPV infection in tissue (Cooper *et al.*, 1991). This approach has also been applied to the study of smear specimens (Troncone *et al.*, 1992; Herrington *et al.*, 1992).

ISH with DIG–labeled probes has also been used for cytomegalovirus (CMV; Gentilomi *et al.*, 1989; Musiani *et al.*, 1990a, Musiani *et al.*, 1990b). An interesting application has been the use of a DIG–CMV probe in double hybridization in combination with a biotin–labeled probe for herpes simplex virus (HSV; Gentilomi *et al.*, 1992). DIG–ISH using DNA probes has been used for detection of human immunodeficiency virus (HIV) in a sensitive single–cell assay of HIV transcription *in vitro* and *in vivo* (Lawrence *et al.*, 1990), for human polyoma virus (JCV) and HPV (Furuta *et al.*, 1990), and Epstein–Barr virus DNA in nasopharyngeal carcinoma (Permeen *et al.*, 1990), where the assay time was reduced to 96 h compared to five weeks. The use of DIG–DNA probes has also contributed to studies of human parvovirus 19 infection in pregnancy (Schwarz *et al.*, 1990; Morey *et al.*, 1992). DIG–labeled oligonucleotides have been a useful tool in study of Coxsackieviruses in mice (Hilton *et al.*, 1992).

One of the latest developments has been the use of DIG for labeling RNA transcripts (Höltke and Kessler, 1990). DIG is incorporated into the nascent RNA probe by SP6, T7, or T3 RNA polymerases, using a cloned DNA fragment, juxtaposed to the promoter for the RNA polymerase, as a template. In comparison with DNA probes the RNA probes have advantages like single strandedness, no self–complementarity, defined length and higher stability of RNA/DNA–hybrids (Höltke and Kessler. 1990). Investigators at the Biochemical Research Center of Boehringer Mannheim have determined the optimal labeling conditions and used the method in dot, Southern and Northern blots.

The first to use a digoxigenin labeled RNA probe for ISH in a virological application were Hukkanen *et al.* (1990), who used it for detection of latency–associated RNA of HSV in a mouse model. In this report, the methodology has been described in an extremely detailed way. An application of DIG–RNA probes has been used in a study of latency–related transcripts (EBERS) of another herpesvirus, EBV (Hamilton–Dutoit *et al.*, 1993). The sensitivity of DIG–RNA probes has been found similar to [^{3}H]–labeled probes, with easy visualization at low magnifications in an ISH study of rabies virus in mouse brains (Jackson, 1992). DIG–RNA probes have been applied to the study of hepatitis C virus in liver tissue (Nouri Aria *et al.*, 1993) and to studies of HIV infection (Wu *et al.*, 1993).

A very powerful tool in study of viral pathogenesis is the combination of PCR and *in situ* hybridization. In an application of this methodology for study of hepatitis C infection, DIG–labeled oligonucleotides or DNA probes were used, as was incorporation of DIG–labeled nucleotide in the PCR product (Nuovo *et al.*, 1993). This approach provides a very high sensitivity, though one must take steps in order to prevent nonspecific DNA synthesis from native human DNA template.

The staining of tissue specimens after detection facilitates recognition of the cell types in the tissue, but the staining may also mask signals from the bound DIG probe, especially if the signal is cytoplasmic (Nouri Aria *et al.*, 1993). This is usually no problem in studies of HPV infection (Heino *et al.*, 1989) or latent HSV infection (Hukkanen *et al.*, 1990), where the signals are confined to the nuclei of infected cells.

B. Dot Blot Hyhridization

Kimpton *et al.* (1989) studied the sensitivity of DIG–probes in a dot–blot system for cytomegalovirus (CMV). Biotinylated probes detected 10–50 pg of homologous DNA, whereas digoxigenin or radioactive label, produced by random priming, detected 0.1 pg of homologous DNA. The method was sensitive enough for detecting CMV in urine specimens. The same sensitivity was found also for HPV 16 (Morris *et al.*, 1990). Furuta *et al.* (1990) were able to detect 1 pg of HPV 11 DNA with dot blot hybridization. Zerbini *et al.* (1990) and Azzi *et al.* (1990) showed the usefulness of a DIG–dot–blot assay for detection of human parvovirus B19 in serum specimens.

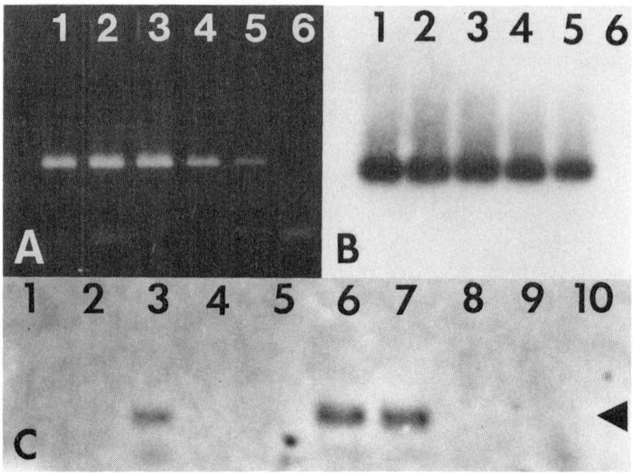

Figure 3.
A). Ethidium bromide staining of PCR products from a test for herpes simplex virus type 1 (HSV–1); (B). Hybridization of the specimens in panel A using a DIG–labeled oligonucleotide probe, visualized by chemiluminescent detection on X–ray film. The lanes are: 1, 10^4 pfu (plaque–forming units) of HSV–1/specimen, 2, 10^3 pfu; 3, 10^2 pfu; 4, 10 pfu: 5, 1 pfu; 6, 0 pfu; (C). Hybridization of PCR products from a test for herpes simplex virus type 2 (HSV–2), visualized on X–ray film by chemiluminescent detection using Lumigen PPD reagent. The lanes 6 and 7 represent positive HSV–2 controls (1 pfu and 1000 pfu, respectively). The specimen in lane 3 is regarded positive, the other lanes represent controls and specimens negative for HSV–2.

Another application using serum specimens was described by Kejian and Bowden (1991), who developed an assay for hepatitis B virus DNA. Their probe detected 0.25 pg of homologous DNA, which is equivalent to 7×10^4 genome copies. In this study some false–positive reactions were seen when an alkaline denaturation method was used.

C. Polymerase Chain reaction (PCR)

The PCR method has found extensive use in all biomedical research. Digoxigenin labeling is beginning to find a place in detection of PCR products as well. Miyamoto *et al.* (1990a) used it for detection of provirus of human T–cell leukemia virus type II (HTLV) and for transformation studies of the same virus in animal cells (Miyamoto *et al.*, I990b). Another group (Griffais *et al.*, 1990) has applied DIG–PCR for detection of Epstein–Barr virus in cultured cells.

Luminescent reagents are emerging as tools for detection of PCR products. Luminescent reagents are applied for detection of DIG–labeled probes in PCR tests for HPV (Chikhaoui *et al.*, 1992), HIV (He *et al.*, 1993) and hepatitis B virus (Yang *et al.*, 1993). The detection of PCR products in a test for HSV using a luminescent detection system (AMPPD, later Lumigen PPD; Boehringer Mannheim) for DIG–labeled probes was equally sensitive as the [^{32}P]–based detection, especially in lower template amounts which are the most critical (Puchhammer–Stoeckl *et al.*, 1992). Our own experience (Hukkanen, unpublished) on application of Lumigen PPD detection of HSV PCR products has been that the DIG label is sufficiently sensitive and extremely rapid, requiring often only 5 min. exposure time on X–ray film (see Fig. 3).

The latest application of PCR is the combination of PCR and *in situ* hybridization (Nuovo *et al.*, 1993) which is discussed above with other *in situ* hybridization studies.

IV. PROTOCOLS FOR DIGOXIGENIN LABELING AND DETECTION OF NUCLEIC ACIDS

In the following paragraphs we describe the essential steps of labeling DNA and RNA probes with digoxigenin and give protocols for detection of DIG–labeled hybrids in tissue sections and on nylon filters. The hybridization and washing steps in individual applications are often very similar to those used with radiolabeled probes, hence they are not described.

A. Labeling of DNA Probes

The digoxigenin labeling method was developed by Boehringer Mannheim Corporation, which also provides the only commercial kits available for DIG–labeling (Genius™ Nonradioactive Nucleic Acid Labeling and Detection System;

DIG Nucleic Acid Detection Kit). Consequently, the reader is referred to the instructions for these kits, where the labeling procedure has been described in detail. Variables and optimal conditions for the method are described by Herrington *et al.* (1989a), Martin *et al.* (1990) and Höltke and Kessler (1990).

Labeled–DNA probes are generated with Klenow polymerase by random–primed incorporation of digoxigenin–labeled deoxyuridine–triphosphate (dUTP). From 10 ng to 3 µg DNA can be labeled in a standard reaction. The linear DNA is added to a microfuge tube in a volume of 15 µl of distilled water. The double–stranded DNA must be denatured by heating DNA at 95 to 100°C for 10 min and chilled immediately after heating in an ice bath. Two µl of the hexanucleotide mixture and 2 µl of the dNTP labeling mixture, both from the DNA Labeling and Detection Kit (see above) are added to the microfuge tube. Finally, 1 µl of Klenow enzyme, provided in the kit, is added. The reaction mixture is incubated at 37°C for 4 h during which 0.5 µl Klenow enzyme is added at 2 h. The reaction is finished by adding 2 µl of 0.2M EDTA, pH 8.0. The labeled DNA is ethanol–precipitated and washed with 70% ethanol, dried and dissolved in 10 mM Tris–HCl, 1 mM EDTA (TE) buffer, pH 8.0.

The reaction kinetics can be followed, if desired, by adding a radioactive labeled tracer, e.g. [^{3}H]–dCTP, which is vacuum–dried at the bottom of the reaction tube. After addition of the Klenow polymerase to the reaction mixture, 0.3 µl aliquots of the mixture can be removed at 1 h intervals onto a filter paper disc. DNA on the filter paper is precipitated first with 10% trichloroacetic acid (TCA) for 10 min and then twice with 5% TCA for 10 min, after which the filter paper is air–dried. The incorporated radioactivity is measured by scintillation spectrometry.

B. Labeling of RNA Probes

The DNA to be transcribed is cloned into the polylinker site of appropriate transcription vectors which contain promoters for SP6 or T7 RNA polymerases adjacent to the polylinker. Labeled RNA probes are synthesized by *in vitro* transcription of DNA with SP6 or T7 RNA polymerases using DIG–labeled UTP as substrate.

The linearized DNA (usually about 1 µg) to be transcribed is added to a microfuge tube on ice in a volume of 13 µl. Two µl of nucleotide triphosphate labeling mixture and and 2 µl of 10X transcription buffer, both provided in the DIG–RNA Labeling Kit SP6/7 (see above) are added. RNase inhibitor (1 µl) and RNA polymerase (2 µl SP6 or T7), both from the kit, are added to the microfuge tube. The reaction is allowed to take place for 2 h at 37°C.

The template DNA can be removed by addition of 2 µl DNase I, RNase free, provided in the kit, and by incubation for 15 min at 37°C. This step, however, is usually not necessary, as the transcription reaction produces a tenfold excess of the DIG–labeled RNA transcript compared with the DNA template. Two µl of 0.2 mM EDTA, pH 8.0, is added to stop the reaction. The labeled RNA is ethanol–

precipitated and the pellet is washed with 70% ethanol, dried and resuspended in diethyl–pyrocarbonate–water or TE–buffer. RNase inhibitor from the kit is added to inhibit possible contaminating RNases. An aliquot of the transcript is analyzed by agarose gel electrophoresis and ethidium bromide staining or by subsequent blotting and incubation of the filter in substrate solution. The kinetics of the labeling reaction can be followed by adding a radioactive labeled tracer NTP, e.g. [α–^{35}S]–CTP and by measuring its incorporation by TCA precipitation and scintillation spectrometry.

C. Detection of Hybrids

In the following, we describe the detection of DIG–labeled nucleic acids in tissue sections on microscope slides and detection of hybrids on a nylon filter using a chemiluminescent substrate.

In both protocols, the nonspecifically bound probes are first washed away after hybridization. The tissue sections or the nylon membrane are then soaked for 2 min in buffer 1 (100 mM Tris, pH 7.5, 150 mM NaCl).

The membrane or slides are soaked in blocking buffer for 30–60 min at 20°C (blocking buffer: 0.5–1.0% blocking reagent in the DIG DNA Labeling and Detection kit, diluted in buffer 1).

The slides or membrane are dipped into buffer 1.

The tissue sections or the membrane are then reacted with dilutions of the anti–digoxigenin antibody:alkaline phosphatase conjugate for 30 min at 20°C (750 mU/ml for tissue sections, 150 mU/ml for nylon membrane, diluted in buffer 1).

The sections or membrane are then washed twice in buffer 1 at 20°C, 15 min per time. Thereafter, they are soaked in buffer 3 for 2 min (buffer 3: 100 mM Tris, pH 9.5, 1000 mM NaCl, 50 mM $MgCl_2$).

For tissue sections, the color solution (4.5 µl of nitroblue tetrazolium solution and 3.5 µl of X–phosphate solution from the DIG DNA Labeling and Detection kit, diluted in 1 ml of buffer 3) is applied on each section and incubated for 5–12 h in a dark, humidified chamber at 20°C. Thereafter, the slides are dipped into TE solution (10 mM Tris, pH 8.0, 1 mM EDTA) and counterstained if necessary. Sometimes counterstaining will mask cytoplasmic signals. Coverslips can be applied and mounted with aqueous mounting medium or Permount™. Slides can be observed while the color development procedes. The nylon membrane is reacted with a dilution of Lumigen PPD stock solution (diluted to 0.1 mg/ml in buffer 3) for 5 min at 20°C in low illumination. The substrate solution can be collected, stored in the dark at 4°C and reused once. The filter is dried for a short time on Whatman 3MM paper and sealed in a hybridization bag. The membrane is incubated at 37°C for 15 min and exposed on Hyperfilm MP (Amersham) for 5–60 min (arbitrary, multiple exposures are recommended). The filter can be reprobed according to the manufacturer's instructions.

V. CONCLUSION

Traditionally, radiolabeled probes are widely used in nucleic acid hybridization protocols. Handling of radioactive materials requires special safety measures and working spaces. Considering this and the short shelf life of radiolabeled probes, the nonisotopic labeling systems have been met with great interest. The most commonly used nonisotopic labeling and detection systems are the biotin–avidin and digoxigenin–anti–digoxigenin system. The DIG–labeled hybrids are detected using an antibody to digoxigenin, which can be visualized by an enzyme–catalyzed color reaction, immunofluorescence or chemi–luminescence. In most studies the digoxigenin labeling system has been found to be as equally sensitive as the radioactive labels and more sensitive than biotinylated probes. The DIG labeling system is very advantageous in applications of *in situ* hybridization due to excellent cellular localization of the signal, without the problem of background staining observed in some tissues when using biotinylated probes. The chemiluminescent detection systems with DIG labeled probes have potential in blot hybridization, opening the possibility for quantitation of the signal. Other advantages of DIG labeling are the stability of the probes, variety of detection methods, and shorter detection times than those obtainable with autoradiography .

REFERENCES

Azzi, A., Zakrzewska, K., Gentilomi, G., Musiani, M., and Zerbini, M. (1990). *J. Virol. Method.* **27,** 125–134.

Chikhaoui, Y., Balaguer, P., Terouanne, B., Boussioux, A.–M., and Nicolas, J.–C. (1992). *J. Immunol. Method.* **154,** 51–56.

Cooper, K., Herrington, C. S., Graham, A. K., Evans, M. F., and McGee, J. O. (1991). *J. Clin. Pathol.* **44,** 400–405.

Furuta, Y., Shinohara, T., Sano, K., Meguro, M., and Nagashima, K. (1990). *J. Clin. Pathol.* **43,** 806–809.

Gentilomi, G., Musiani, M., Zerbini, M., Gallinella, G., Gibellini, D., and La Placa, M. (1989). *J. Immunol. Method.* **125,** 177–183.

Gentilomi, G., Musiani, M., Zerbini, M., Gibellini, G., Gallinella, G., and Venturoli, S. (1992). *J. Histochem. Cytochem.* **40,** 421–425.

Griffais, R., André, P. M., and Thibon, M. (1990). *Res. Virol.* **141,** 33 I–335.

Hamilton–Dutoit, S. I., Raphael, M., Audouin, J., Diebold, I., Lisse, I., Pedersen, C., Oksenhendler, E., Marelle, L., and Pallesen, G. (1993). *Blood* 82, 619–624.

He, Y., Coutlee, F., Saint–Antoine, P., Olivier, C., Voyer, H., and Kessous–Elbaz, A. (1993). *J. Clin. Microbiol.* **31,** 1040–1047.

Heiles, H. B. J., Genersch, E., Kessler, C., Neumann, R., and Eggers, H. J. (1988). *BioTechniques* **6,** 978–981.

Heino, P., and Hukkanen, V. (1992). In *Nonradioactive labeling and detection of biomolecules.* (C. Kessler, ed.), pp. 316–324. Springer–Verlag, Berlin and Heidelberg.

Heino, P., Hukkanen, V., and Arstila, P. (1989). *J. Virol. Method.* **26,** 331–338.

Herrington, C. S., Burns, J., Graham, A. K., Evans, M., and McGee, J. O. (1989a). *J. Clin. Pathol.* **42,** 592–600.

Herrington, C. S., Bums, J., Graham, A. K., Bhatt, B., and McGee, I. O. (1989b). *J. Clin. Pathol.* **42,** 601–606.

Herrington, C. S., deAngelis, M., Evans, M. F., Troncone, G., and McGee, J. O. (1992). *J. Clin. Pathol.* **45,** 385–390.

Hilton, D. A., Day, C., Pringle, J. H., Fletcher, A., and Chambers, S. (1992). *J. Virol. Method.* **40,** I55–162.

Hukkanen, V., Heino, P., Sears, A., and Roizman, B. (1990). *Method. Mol. Cell. Biol.* **2,** 70–81.

Höltke, H.–J., and Kessler, C. (1990). *Nucleic Acids Res.* **18,** 5843– 5851.

Jackson, A. (1992). *Mol. Cell. Prohes* **6,** 131–136.

Kejian, G., and Bowden, D. S. (1991). *J. Clin. Microhiol.* **29,** 506–509.

Kimpton, C. P., Corbitt, G., and Morris, D. J. (1989). *J. Virol. Method.* **24,** 335–346.

Konno, R., Shikano, K., Horiguchi, M., Endo, A., Chiba, H., Yaegashi, N., Sato, S., Yajima, H., Tase, T., and Yajima, A. (1990). *Tohoku J. Exptl. Med.* **160,** 383–390.

Langer, P. R., Waldrop, A. A., and Ward, D. C. (1981). *Proc. Natl. Acad. Sci. USA* **75,** 6633–6637.

Lawrence, J. B., Marselle, L. M., Byron, K. S., Johnson, C. V., Sullivan, J. L., and Singer, R. H. (1990). *Proc. Natl. Acad. Sci. USA* **87,** 5420– 5424.

Lion, T., and Haas, O. A. (1990). *Anal. Biochem.* **188,** 335–337.

Martin, R., Hoover, C., Grimme, S., Grogan, C., Höltke, J., and Kessler, C. (1990). *BioTechniques* **9,** 762–765.

Miyamoto, K., Tomita, N., Ohtsuki, Y., and Kitajima, K. (1990a). *Jap. J. Cancer Res.* **81,** 313–316.

Miyamoto, K., Tomita, N., Hayashi, K., and Akagi, T. (1990b). *Jap. J. Cancer Res.* **81,** 720–722.

Morey, A. L., Porter, H. J., Keeling, J. W., and Fleming, K. A. (1992). *J. Clin. Pathol.* **45,** 673–678.

Morris, R. G., Arends, M. J., Bishop, P. E., Sizer, K., Duvall, E., and Bird, C. C. (1990). *J. Clin. Pathol.* **43,** 800–805.

Musiani, M., Gentilomi, G., Zerbini, M., Gibellini, D., Gallinella, G., Pileri, S., Baglioni, P., and La Placa, M. (1990a). *Histochemistry* **94,** 21–25.

Musiani, M., Zerbini, M., Gentilomi, G., Gallinella, G., Venturoli, S., Gibellini, D. , and La Placa, M. (1990b). *J. Clin. Microhiol.* **28,** 2101– 2103.

Nouri Aria, K. T., Sallie, R., Sangar, D., Alexander, G. J. M., Smith, H., Byrne, J., Portmann, B., Eddleston, A. L. W. F., and Williams, R. (1993). *J. Clin. Invest.* **91,** 2226–2234.

Nuovo, G. I., Lidonnici, K., MacConnell, P., and Lane, B. (1993). *Am. J. Surg. Pathol.* **17,** 683–690.

Permeen, A. M. Y., Sam, C. K., Pathmanathan, R., Prasad, U., and Wolf, H. (1990). *J. Virol. Method.* **27,** 261–265.

Puchhammer-Stoeckl, E., Heinz, F. X., and Kunz, C. (1992). *J. Virol. Method.* **43,** 257–266.
Schwarz, T. F., Nerlich, A., and Roggendorf, M. (1990). *Behring Inst. Mitt.* **85,** 69–73.
Stürzl, M., Oskoui, K. B., and Roth, W. K. (1992). *Mol. Cell. Probes* **6,** 107–I 14.
Troncone, G., Herrington, C. S., Cooper, K., deAngelis, M. L., and McGee, I. OD. (1992). *J. Clin. Pathol.* **45,** 305–313.
Wu, T.-C., Kanayama, M. D., Hruban, R. H., Whitehead, W., and Raj, N. B. K. (1993). *Am. J. Pathol.* **142,** 25–31.
Yang, G., Ulrich, P., Aiyer, R., Rawal, B., and Vyas, G. (1993). *Blood* **81,** 1083–1085.
Zakrzewska, K., Ciappi, S., and Azzi, A. (1993). *Mol. Cell. Probes* **7,** 55– 60.
Zerbini, M., Musiani, M., Venturoli, S., Gallinella, G., Gibellini, D., Gentilomi, G., and LaPlaca, M. (1990). *J. Clin. Microbiol.* **28,** 2496– 2499.

8

Biospecific Interaction Analysis in Real Time Using a Biosensor System with Surface Plasmon Resonance Detection

Lars G. Fägerstam

I. INTRODUCTION

All biological molecules express their function through interactions with other molecules. Characterization of these interactions is therefore of fundamental interest in many areas of biological research and biotechnological development. A promising area of development in biospecific interaction analysis is the use of biosensors to detect and measure interactions in real time. A biosensor may be defined as an instrument that combines a biological recognition mechanism with a sensing device or transducer (Ngwainbi, 1990).The transducer generates a measurable signal in response to the specific recognition process. This chapter describes a system using flow injection technology combined with a biosensor based on the optical phenomenon surface plasmon resonance (SPR) for detection of biospecific interactions. SPR detection as described here allows monitoring of the interactions as they occur. SPR has previously been applied to concentration analysis of biomolecules (Liedberg *et al.*, 1983) and for monitoring of immunocomplex formation (Flanagan *et al.*, 1984, Cullen *et al.*, 1987, Daniels *et al.*, 1988, Kooyman *et al.*, 1988).

II. SURFACE PLASMON RESONANCE DETECTION PRINCIPLE

Surface plasmon resonance is an optical phenomenon arising in connection with total internal reflection of light at a metal film–liquid interface. Normally, light traveling through an optically denser medium is totally reflected back into this medium when reaching an interface to an optically less dense medium

Lars G. Fägerstam • Pharmacia Biosensor AB, S–751 82 Uppsala, Sweden

Applied Virology Research, Vol. 3, Edited by
E. Kurstak *et al.*, Plenum Press, New York, 1994

provided the angle of incidence is larger than the critical angle. This is known as total internal reflection. Although the light is totally reflected, a component of the incident light momentum called the evanescent wave penetrates a distance of the order of one wave length into the less dense medium. The evanescent wave may be used to excite molecules close to the interface, as in total internal reflection fluorescence (Kronick and Little, 1975). If however the light is monochromatic and polarized and the interface between the media is coated with a thin (a fraction of the light wave length) metal film, the evanescent wave under certain conditions will interact with free oscillating electrons (plasmons) in the metal film surface. When surface plasmon resonance occurs light energy is dissipated across the interface between the two media and the reflected light intensity is thus decreased. The resonance phenomenon will only occur for light incident at a sharply defined angle which, when all else is kept constant, is dependent on the refractive index of the less dense medium. Changes in the refractive index out to about 300 nm from the metal film surface can thus be followed by continuous monitoring of the resonance angle. A detection volume is defined by the size of the illuminated area at the interface and the penetration depth of the evanescent field. It is noteworthy that no light passes through the detection volume. The optical device at one side of the metal film detects changes in the refractive index in the medium at the other side. Extensive theoretical treatment of the SPR phenomenon and reviews of experimental results have been published previously (Boardman, 1982, Raether, 1988).

III. INSTRUMENTATION

The instrument, BIAcore™, described here was developed by Pharmacia Biosensor AB, Uppsala, Sweden. It consists of a processing unit, reagents for ligand immobilization, exchangeable sensor chips and a personal computer for control and evaluation. The processing unit contains the SPR monitor and an integrated microfluidic cartridge that, together with an auto sampler, controls delivery of sample plugs into a transport buffer that continuously passes over the sensor chip surface.

In Fig. 1 is shown the configuration of the SPR detector, the sensor chip and the microfluidic cartridge. The sensor chip is held in contact with the prism of the optical system by the microfluidic cartridge. Plane polarized light from a high efficiency near–infrared LED is focused in a transverse wedge, through the prism onto the side of the sensor chip opposite to the gold layer. Reflected light is monitored by a fixed two dimensional array of light sensitive diodes, placed so that the resolution between diodes correspond to a difference in reflected angle of 0.1°. Computer interpolation routines process the data from the diode array to determine the position of the resonance angle to an accuracy of 10^{-4}°. The use of a fixed detector array eliminates moving parts from the optical system, increasing reliability of the instrument and allowing changes in resonance angle to be followed in real time. A layer of silicone polymer, matched in refractive

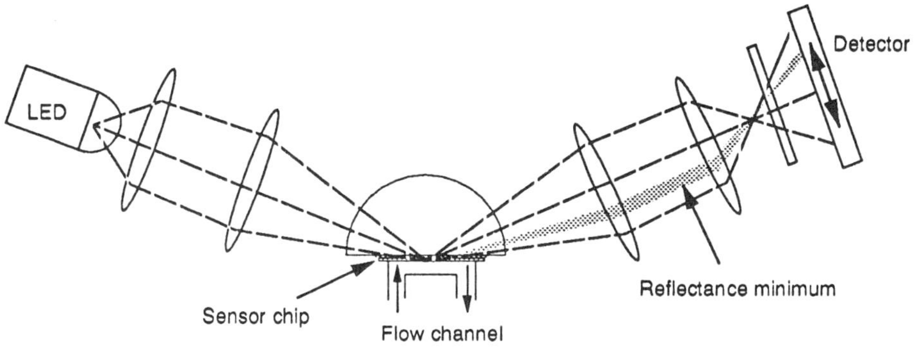

Figure 1.
The SPR detector, sensor chip and liquid flow cell are shown docked together. Light from the LED is coupled to the sensor chip via a prism. The position of the reflectance minimum is monitored continuously by a photodiode array. When a molecule introduced via the flow cell is captured at the sensor surface, the position of the reflectance minimum will change. In the instrument, a wedge-shaped light beam and a two-dimensional detector array enables monitoring of four flow cells.

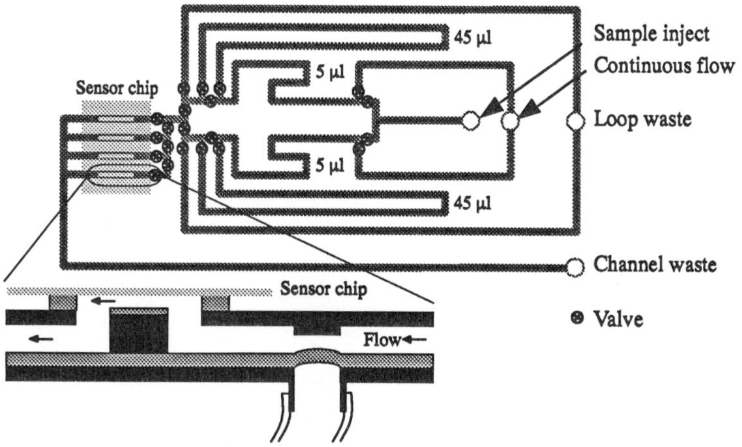

Figure 2.
The microfluidic cartridge contains pneumatically activated microvalves that control the sample loops and choice of flow cell. The sensor chip is docked directly onto the cartridge, minimizing the distance between sample loops and detector flow cells.

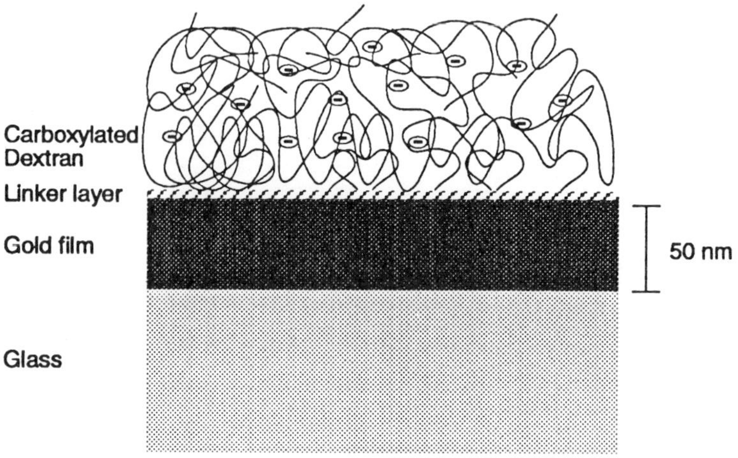

Figure 3.
The sensor chip consists of a thin gold film on a glass support. A matrix of carboxymethylated dextran is bonded to the gold surface via a linker layer. The dextran matrix, which extends about 100 nm out from the surface, serves as a support to which the first interactant is immobilized.

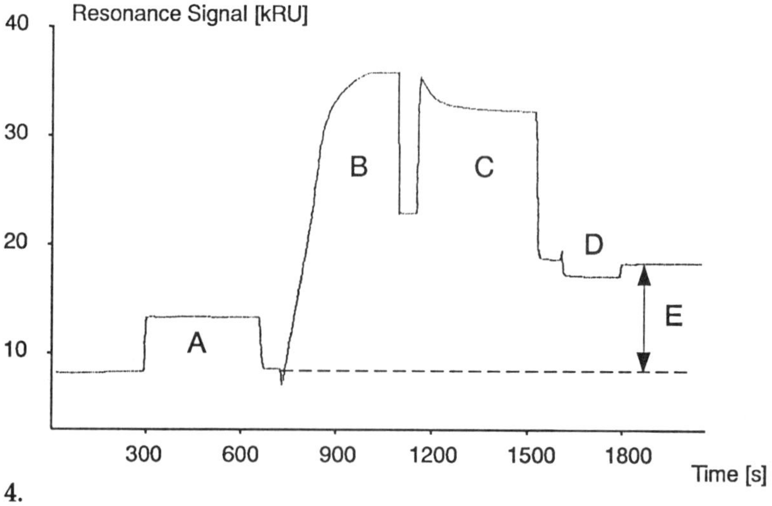

Figure 4.
Sensorgram of all steps during immobilization of rabbit anti–mouse Fc (RAMFc) to the sensor chip. A buffer flow of 5 ml/min continuously passes over the sensor surface. A 30 ml pulse of activation chemicals (A) is injected followed by 30 ml of a 30 mg/ml solution of RAMFc (B). A deactivation reagent (C) is then injected. The RAMFc–surface is then conditioned with 15 ml of 100 mM HCl (D). The amount of RAMFc immobilized, 10.5 kRU corresponding to about 10.5 ng/mm^2, is indicated at (E).

index to the glass of the sensor chip, ensures exact optical contact between the removable sensor chip and the fixed parts of the optical system.

The microfluidic cartridge (Fig. 2.) contains two identical sets of valves and channels, each with two sample loops for injection of volumes up to 5 and 45 microliters, respectively. At the surface of the cartridge a flow cell block with four channels is located. When the cartridge is docked against the sensor chip four parallel flow cells, each with a volume of 60 nanoliters, are formed. The four flow cells, which are illuminated simultaneously by the transverse wedge of light, can be connected to any of the sample loops through the valves.

The sensor chip (Fig. 3) consists of a glass substrate onto which a 50 nm thick gold film has been deposited. The gold film is then covered with a long chain hydroxyalkyl thiol which forms a monolayer at the surface (Nuzzo and Allara, 1983, Bain *et al.*, 1989). This layer serves both as a barrier to prevent proteins and other ligands from coming into contact with the metal and as a functionalized structure to which a matrix of dextran is attached. The dextran which extends typically 100 nm out from the surface makes it possible to exploit the penetration depth of the evanescent field creating a volume in which molecular interactions can be studied. By using a carboxymethylated dextran, substances containing primary amine functions can be immobilized after activation of the matrix with carbodiimide/N–hydroxysuccinimide (Löfås and Johnsson, 1990).

IV. PREPARATION OF SENSOR SURFACES

The first experimental step is to immobilize a ligand to the sensor surface. The ligand may be one of the reactants in the interaction of interest, or e.g. an antibody that can capture one of the reactants. The immobilization, which is performed with the sensor chip in situ in the instrument, is continuously monitored by the SPR detector (Fig. 4). This allows direct comparison of the amount immobilized between all surfaces prepared. Typically 30 ml of a solution containing 10 to 100 mg/ml of the ligand is required for the immobilization. Depending on the stability of the immobilized ligand, the sensor surface can be regenerated and used for a series of analyses. When a monoclonal antibody is used as the immobilized ligand typically 50–100 cycles can be run without loss of activity.

V. THE SENSORGRAM

By continuously monitoring the refractive index (RI) in the detected volume, and plotting this value against time, a *sensorgram* is obtained. The abscissa of the sensorgram is denoted the *resonance signal* and is indicated in *resonance units (RU)*. 1000 RU corresponds to a 0.1° shift in the surface plasmon resonance angle and for the average protein this corresponds to a surface concentration change of about 1 ng/mm^2 (Stenberg *et al.*, 1991). The total range covered by the SPR detector is 3° corresponding to 30,000 RU. The resonance signal at a certain point in time will be the sum of the contributions from the sensor surface, the interacting molecules and the bulk solution. Under conditions of constant bulk refractive index, the amount of interacting molecules can be monitored continuously. Otherwise, the amount may be quantified by readings taken between sample injections, where the transport buffer of constant RI passes the sensor surface.

Let us consider the sensorgram shown in Fig. 5. The sensor surface used has been prepared by immobilizing a polyclonal rabbit anti–mouse Fc (RAMFc) antibody as shown in Fig. 4. At the beginning of the run a continuous flow of buffer is passing the surface (A), resulting in a resonance signal of 20,350 RU. During the time span from (A) to (B) a crude cell culture supernatant, containing a mouse monoclonal antibody (MAb), is introduced by switching one of the sample loops into the buffer flow. The instantaneous rise in the resonance signal as the sample enters the flow cell is due to the high refractive index of the culture supernatant. At (B) the sample plug has passed and buffer once again passes the surface. Here a resonance signal level of 21,550 RU can be read under the same conditions as in (A). The signal shift from binding of the MAb to the RAMFc surface is thus 1,200 RU. During time (B) to (C) the antigen, corresponding to the captured MAb, is introduced. The signal gradually increases another 250 RU. When buffer replaces the antigen solution (C), the dissociation of the MAb–antigen complex can be seen. By introducing a regenerating agent (D) the MAb is washed out and the signal drops back to its original level. The RAMFc surface is now ready for another analysis cycle (E). This illustrates several features of the technique: 1) the selectivity of the sensor surface can be controlled by the operator through the immobilized molecule; 2) concentration determinations in crude samples can bé made if a standard curve is constructed; 3) different unlabeled molecules can be injected in series and each interaction quantified, allowing stoichiometric calculation on the complex formed; 4) the progress of association and dissociation events can be followed, providing a basis for kinetic studies and 5) the sensor surface can be regenerated as many times as the immobilized ligand allows.

VI. APPLICATIONS

A. Binding site studies

Since SPR detection requires no labelling of the interacting molecules and allows quantification of each interactant in the sequential formation of multimolecular complexes, binding site studies is an obvious application of the technique. Binding site studies are of importance for example in the selection of monoclonal antibodies during development of diagnostic tests and in the elucidation of functionally important structural elements e.g. in receptor–ligand interaction studies.

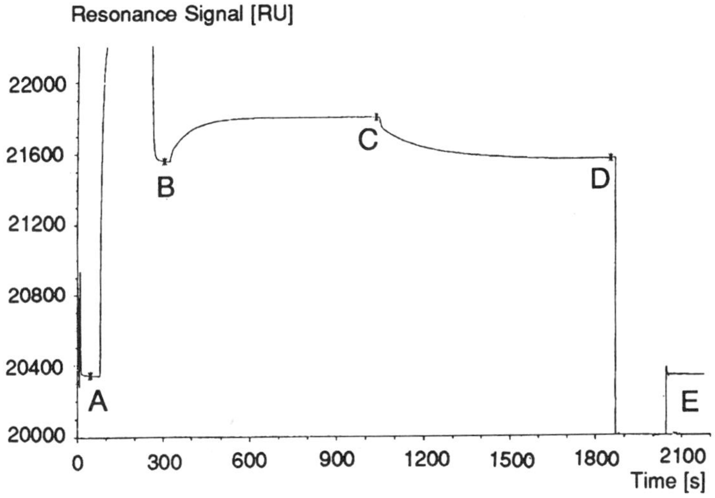

Figure 5.
Sensor gram of a representative analysis cycle in the biosensor. At (A) the continuous buffer passes a RAMFc surface prepared as described in Fig. 4. Between (A) and (B) a pulse of crude hybridoma cell culture supernatant containing a MAb is introduced. Next, the antigen is injected and the rate of antibody–antigen complex formation can be followed. When the antigen solution is replaced by buffer (C), the complex starts to dissociate. At (D) the experiment is terminated by injection of a regenerating agent that removes the all material down to RAMFc. The RAMFc surface is now ready for the next analysis cycle (E).

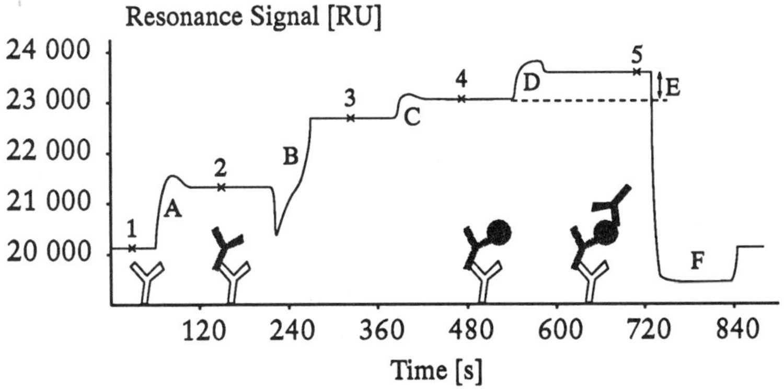

Figure 6.
Sensorgram of a two–site binding assay cycle. At point 1 a resonance signal reading is registered as buffer passes the RAMG1 surface at 5 ml/min. 4 ml of each of the first MAb (A), blocking MAb (B), antigen p24 (C) and the second MAb (D) are then injected in sequence. At the indicated times (2–5) successive resonance signal readings are registered. The response from binding of the second antibody is indicated at (E). At (F) the RAMG1 surface is regenerated with 100 mM HCl and is then ready for the next cycle.

B. Epitope mapping

A number of methods for binding pattern studies or epitope mapping of MAbs have been described in the literature. By testing the ability of all possible pairs of MAbs in a two–site or double antibody binding assay, each MAb can be assigned a specific reaction pattern relative to the other MAbs. All MAbs showing the same binding pattern are then assigned to the same epitope. The resolution in the epitope map is dependent on the number of MAbs available, since each additional MAb might reveal an new pattern from those already tested. Gel filtration high performance liquid chromatography (Crawford et al., 1982, Wilson and Smith, 1984, Mazza and Retegui, 1989), gel electrophoresis (Mazza and Retegui, 1989) and quasi–elastic light scattering spectroscopy (Yarmush *et al.*, 1987) have been used to determine, from the size of the immune complex formed, the ability of paired MAbs to bind to the antigen. Most commonly however, RIA or ELISA methods have been employed. These require labelling of the MAbs to be investigated or the use of a labelled detection antibody directed towards the MAbs. For control experiments it might also be necessary to label the antigen.

In the example described here (Fägerstam *et al.*, 1990) BIAcore was used to elucidate the binding pattern of a panel of monoclonal antibodies (MAbs) to the human immunodeficiency virus type 1 (HIV–1) core protein p24 in a two–site binding assay using the sample injection sequence first MAb–antigen–second MAb. All of the mouse MAbs used in the example were of the isotype IgG1. Therefore a polyclonal rabbit anti–mouse IgG1 (RAMG1) was immobilized to the dextran matrix to serve as a capturing agent for the first MAb that was injected, without previous purification, as crude hybridoma culture supernatants. The use of a capturing agent, instead of direct immobilization of the first MAb has several advantages; i) the MAbs, which might be a minor constituent of the culture media, need not be purified or enriched prior to use. The RAMG1 surface will serve as a specific affinity support, ii) the MAbs are not exposed to covalently modifying chemical agents, iii) fresh MAb is used in each assay cycle since the regeneration removes all material down to the covalently immobilized RAMG1 and iv) different MAbs may be tested in any order wanted using the same sensor chip.

A sensorgram of a two–site binding assay cycle is shown in Fig. 6. After injection of the first MAb, remaining free RAMG1 is blocked by injection of a non–specific IgG1 MAb. The blocking is necessary to assure that the resonance signal from the second MAb emanates from binding to the antigen and not to the RAMG1 surface. The antigen is then injected followed by the second MAb. After regeneration, the RAMG1–surface is then ready for the next analysis cycle.

For 30 MAbs, as used in the example shown, 900 assay cycles would be necessary to test all possible combinations in a two–site binding study. In practice, however, once a particular pair of MAbs tested is found to bind simultaneously to the antigen, it is assumed that they bind to non–overlapping sites and it is not necessary to test them in reverse order. Negative results, however, might be due to effects that depend on the order of binding and can

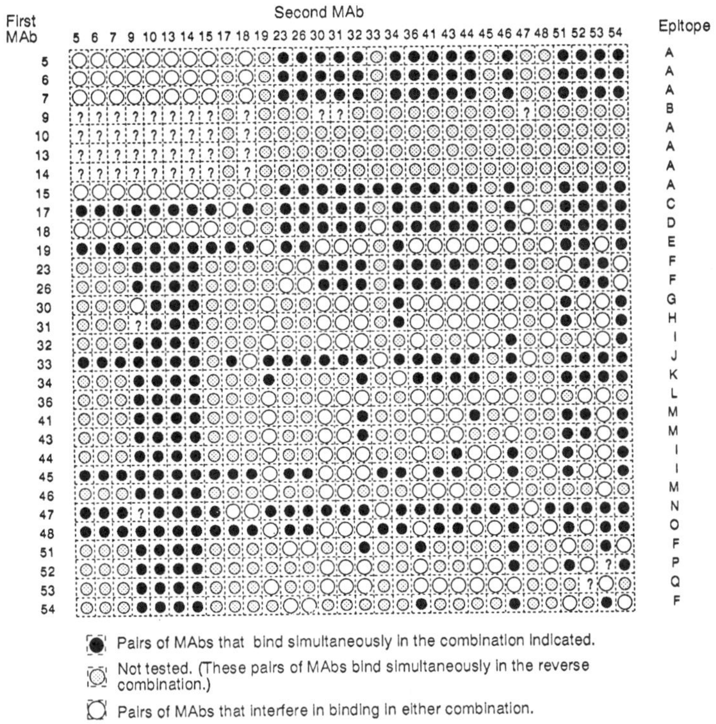

Figure 7.
The reactivity pattern matrix obtained by two–site binding assays. Question marks indicate uncertain interpretation or, for MAbs 9–14, assays that could not be performed due to loss of antigen binding activity when captured by the RAMG1 surface.

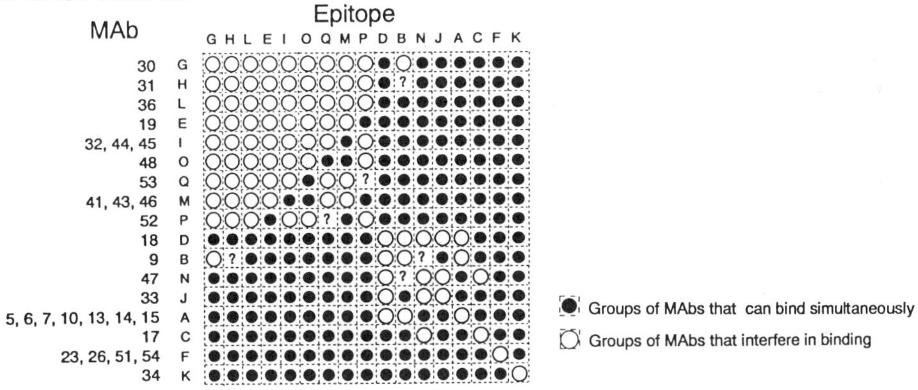

Figure 8.
By grouping MAbs that show the same reaction pattern 17 groups were distinguished. The groups are shown sorted in descending order of binding interference. Note the large cluster of interference att the top left corner and the smaller cluster further down the diagonal. The two clusters are bridged by group B (MAb 9) that interferes with groups G and D, the most dominant epitope groups of the two clusters.

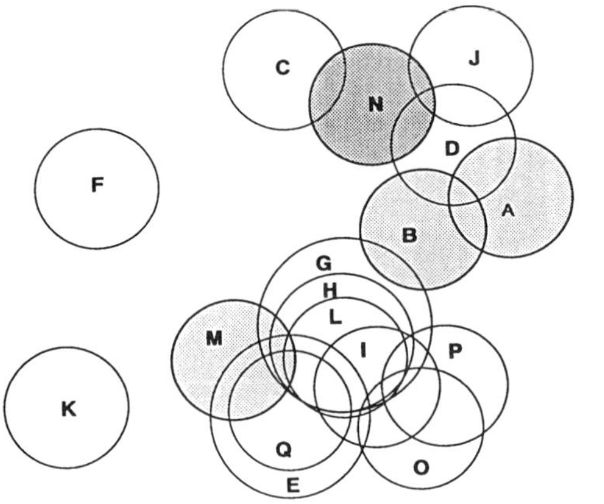

Figure 9.
From the matrix in Fig. 8 a two-dimensional 'surface like' map can be constructed. Overlapping circles denote groups that cannot bind concurrently. Shaded circles indicate groups containing MAbs that could be inhibited by synthetic peptides.

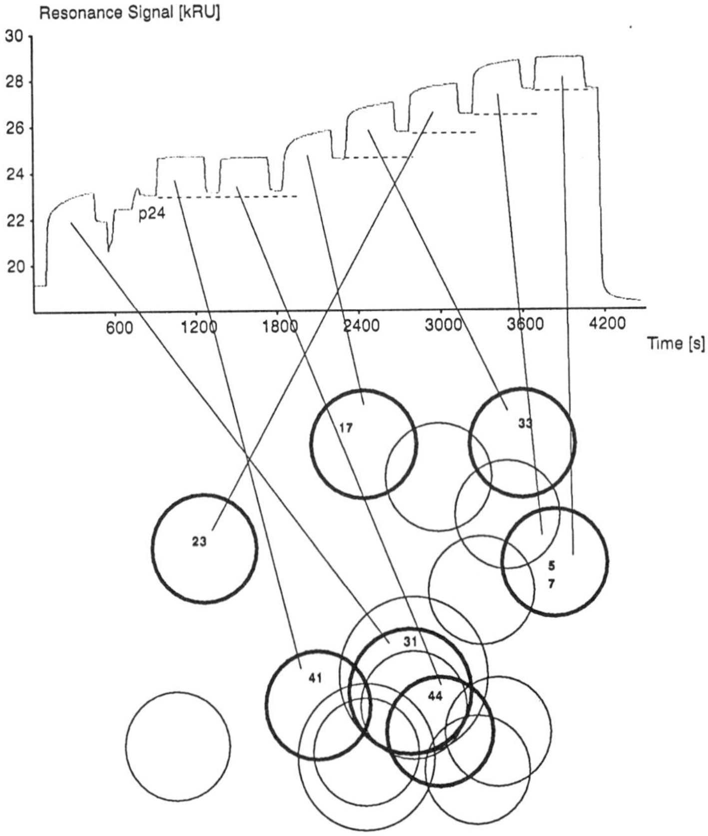

Figure 10.
Sensorgram of a multideterminant binding experiment with eight hybridoma supernatants. The dotted lines indicate the signal level before and after each MAb injection. A raise in the signal level is only obtained for injection of MAbs belonging to groups that do not interfere with any of the previously injected. The final hexamolecular complex thus was formed in the sequence MAb31–antigen p24–MAb17–MAb33–MAb23–MAb5.

only be interpreted after after testing both orders. Since SPR detection allows monitoring of all reactants, the reason for absence of a signal for the second MAb can be looked into in detail. For example, for four of the MAbs (numbers 9, 10, 13 and 14 in Fig. 7) it could be seen from the sensorgrams that the ability to bind antigen was unexpectedly lost when they were captured by RAMG1.

After completion of the two–site binding experiments data are transferred to a reactivity pattern matrix (Fig. 7). By grouping the MAbs, based on their reactivity patterns, 17 different epitope groups were resolved. Fig. 8 shows the binding interference pattern of the 17 groups. For clarity, the groups were sorted in descending order of interference. Note the emergence of two clusters of epitopes, the most prominent at the upper left hand corner of the diagram and a smaller one further down the diagonal. The only overlap between the two clusters is MAb 9 of epitope B which interferes with MAbs from both groups G and D, the most dominant epitope groups of the two clusters. Transformation of the pattern into a two–dimensional diagram (Fig. 9) gives a surface–like representation of the binding interference between the different MAb groups. The diagram should not be interpreted as physical locations of the epitopes on the surface of the antigen since influences such as allosteric conformational changes in the antigen or electrostatic forces between MAbs might distort the pattern.

A more stringent definition of the term 'epitope mapping' used by some authors includes the precise determination of the structures to which the MAbs bind. Methods used to localize protein epitopes involve cross reactivity studies with protein fragments (Atassi, 1984) or synthetic peptides (Geysen *et al.*, 1988), comparison of reactivity to closely related proteins (Benjamin *et al.*, 1984) or to intact proteins chemically modified at single residues (Burnens *et al.*, 1987). Attempts have also been made to predict the location of continuous epitopes in proteins from their primary structure (Van Regenmortel and de Marcillac, 1988).

One approach to such studies that was performed in the epitope mapping study described here was to test the ability of synthetic peptides, derived from the amino acid sequence of the antigen, to inhibit the MAb–antigen interaction. The selection of peptides to be synthesized was done using the prediction method of Hopp and Woods (Hopp and Woods, 1981). After incubation of the MAb–peptide mixtures, they were injected over a sensor surface to which the antigen had been immobilized. The responses obtained were compared to that of peptide free control samples. The immobilization did not abolish the antigen binding ability of any of the MAbs. In Fig. 9 the shaded circles indicate the four epitope groups that showed reactivity to the seven peptides tested. Here again a pattern evolves which resembles the grouping of MAbs by two–site binding. All, but only, the members of group M were inhibited by the same peptide and only the single member of group N was inhibited by another. A more complex pattern was obtained for the MAbs belonging to groups A and B. All of these MAbs showed reactivity towards a 24 amino acid residues long peptide containing a disulfide bond bridging 19 residues forming a loop. When assayed using a 18 amino acid residues long peptide corresponding to the inner part of the loop only two MAbs, one from each of groups A and B where inhibited. For the MAbs showing reactivity towards the peptides tested these results indicate that the grouping deduced from the two-site binding patterns in fact represent groups of MAbs recognizing neighboring structural motifs in the antigen.

Multideterminant binding experiments using labelling methods would require an elaborate experimental protocol. One approach which has obvious practical limitations is to use different labels for different antibodies and to quantify them by discriminative detection. Another approach would be to monitor the stepwise buildup of the multimolecular complex to successively larger forms by detection of a labelled antibody after pre-incubation with different combinations of unlabeled antibodies. Since no labelling is needed when using SPR detection, multideterminant binding experiments can be performed simply by sequential injection of the hybridoma supernatants whilst monitoring the stepwise formation of the antigen-antibody complex. Fig. 10 shows an example of the sequential injection of seven MAbs that together with the first antigen capturing MAb represent five different reactivity pattern groups.

VII. AFFINITY AND KINETIC MEASUREMENTS

A variety of methods is available for measuring affinity (Stanley et al., 1983, Steward, 1986) of biomolecular interactions. In general, the reaction is allowed to reach equilibrium, free and bound reactants are separated and at least one of them is quantified. Once the amount of free or bound reactant is known, equilibrium data can be plotted according to a number of algorithms (Andrade, 1986, Azimzadeh and Van Regenmortel, 1990), providing information not only about the affinity constants but also about the number and possible variety of binding sites. Methods that detect changes in optical parameters like fluorescence (Dandliker and Levison, 1967) or absorbance (Froese, 1968) in one of the reacting species can in certain cases be employed for direct kinetic analysis using native molecules. Kinetic information can also be obtained using methods where it is possible to quickly separate bound and free reactants (Skubitz and Smith, 1975, Mason and Williams, 1986,, Olson *et al.*, 1989, Johnstone *et al.*, 1990). In these cases, however, one of the reactants is often labelled and native molecules are no longer studied.

A. Kinetic analysis

For a one to one interaction in solution between reactants A and B that can form the complex AB, the rate of formation of AB-complexes at time t may be written as:

$$d[AB]/dt = k_{ass}\,[A]_t\,[B]_t - k_{diss}\,[AB]_t$$

where k_{ass} is the association rate constant (unit: $M^{-1}\ s^{-1}$) and k_{diss} is the dissociation rate constant (unit: s^{-1}), or by substituting $[B]_t$ for $[B]_0 - [AB]_t$, where $[B]_0$ is the total concentration of reactant B, as:

$$d[AB]/dt = k_{ass}\,[A]_t\,([B]_0 - [AB]_t) - k_{diss}\,[AB]_t$$

In the biosensor, one of the reactants is immobilized to the sensor surface and the other is continuously replenished from a solution flowing past the sensor surface. The response R will correspond to the amount of AB–complexes formed and the maximum response R_{max} will be proportional to the surface concentration of the immobilized ligand. The rate equation can thus be rewritten as:

$$dR/dt = k_{ass}\,C\,(R_{max} - R_t) - k_{diss}\,R_t$$

where dR/dt is the rate of formation of surface complexes, i.e. the derivative of the observed response curve. C, which is kept constant, is the concentration of analyte in free solution, R_{max} is the total amount of binding sites of the immobilized ligand expressed as SPR response and $(R_{max} - R_t)$ is the amount of remaining free binding sites at time t. Note that since terms in R occur on both sides of the equation, the response value R_t can be used directly without conversion between response and absolute concentration of complexes formed on the sensor chip surface.

Rearranging the rate equation shows more clearly that the derivative of the binding curve is linearly related to the response. In principle, therefore, k_{ass} and k_{diss} can be calculated from the straight line obtained by plotting of dR/dt versus R.

$$dR/dt = k_{ass}\,C\,R_{max} - (k_{ass}\,C + k_{diss})\,R_t$$

This requires knowledge of the saturation response R_{max} which would require injection of the free analyte at very high concentrations. This approach is therefore often unrealistic, especially if precious samples are handled.

From the rearranged rate equation, it can be seen that the slope $(k_{ass}\,C + k_{diss})$ of the line is itself linearly related to the concentration C of free analyte. If therefore the slopes of the dR/dt vs R lines are plotted as a function of analyte concentration C, a new line is obtained with slope k_{ass} and intercept on the abscissa k_{diss}.

A typical example on kinetic analysis (from reference 36) is shown in Figs. 11 a–c. Here the interaction between a MAb and its antigen, HIV–1 core protein p24, is studied. The runs were performed as described in Fig. 5. A RAMFc surface is used to capture a constant amount of MAb from the crude hybridoma culture supernatant at the beginning of each analysis cycle. A series of cycles are run using antigen concentrations in the range 12.5 to 300 nM (Fig. 11a). The slope of the binding rate versus response (Fig. 11b) for each concentration is then plotted against concentration (Fig. 11c). In this particular case the association rate constant k_{ass} was calculated to be $1.4 \times 10^5\ M^{-1}\,s^{-1}$.

A more practical approach to the ranking of the properties of e.g. different monoclonal antibodies is a direct visual comparison of their interaction behavior under standardized conditions. In Fig. 12. which is an overlay of sensorgrams

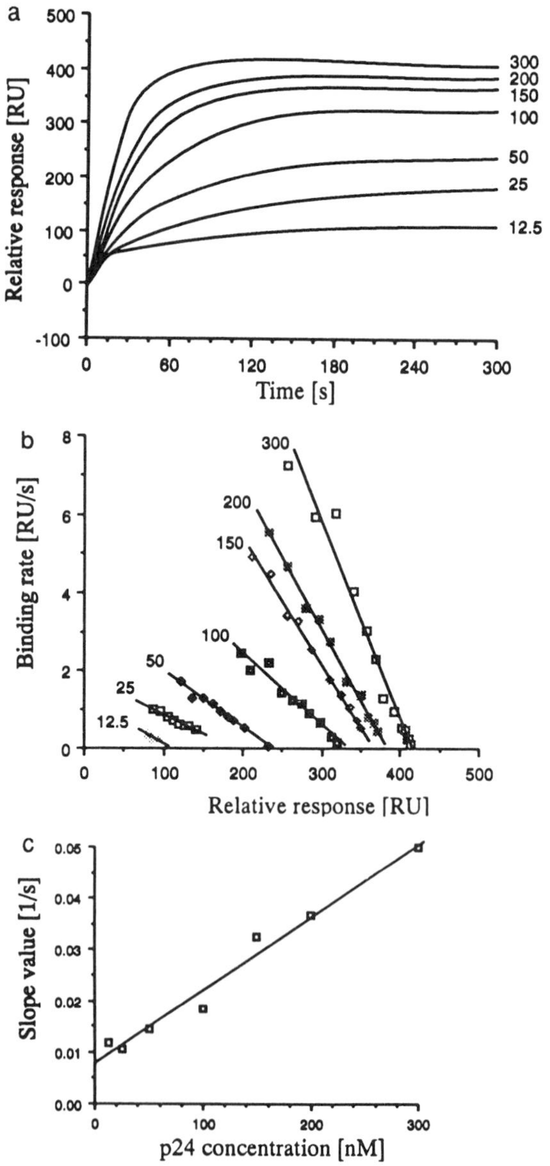

Figure 11.

a Overlay of sensorgrams of the interaction between a MAb captured by a RAMFc surface and its antigen HIV–1 core protein p24. The antigen was injected at seven different concentrations in the range 12.5 to 300 nM.

b Plots of the binding rates versus relative response obtained for the different concentrations.

c From a plot of the slopes of the lines in b) versus the concentrations at which they were obtained, the association rate constant k_{ass} was calculated to be 1.4 X $10^5 M^{-1} s^{-1}$.

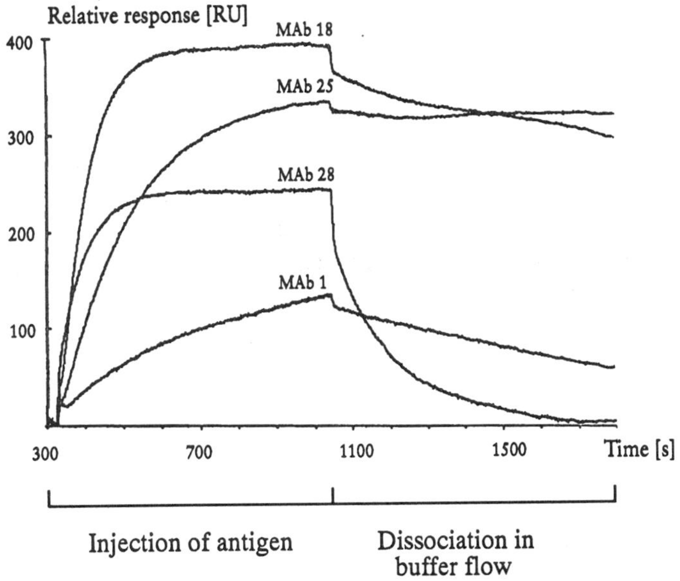

Figure 12.
Overlay of sensorgrams obtained by capturing the same amount of four different MAbs specific for the same antigen.Since the antigen was injected at the same concentration in the four experiments, properties in terms of binding rates, steady state levels and dissociation rates can be directly compared.

from four experiments, equal amounts of four different monoclonal antibodies were captured at a RAMFc surface. The antigen was then injected at a fixed concentration. As can be seen, the four MAbs differ both in their binding rates, steady state levels and dissociation rates. MAbs 18 and 28 reach their binding equilibria after about 300 seconds with MAb 18 showing a higher steady state level, indicating a higher affinity. MAb 25 shows a much slower binding rate and approaches steady state at the end of the injection whilst MAb 1 still shows a significant net association at the end of the injection. From the dissociation part of the sensorgrams it can be seen that MAb 28 releases the antigen at a rate obviously much higher than that of any of the other MAbs. Bearing in mind the fast association of this MAb and that the buffer volume that passes the flow cell during the 800 seconds of dissociation corresponds to about 1100 cell volumes, the affinity constant of the MAb could still be 10^6 M^{-1} or higher. From a practical point of view, this MAb would be a good candidate to use in an immunoaffinity support for chromatographic purification purposes. MAb 18 which shows a rapid and relatively stable binding at a high level would probably be useful as the capturing antibody in a RIA or ELISA type of assay.

This chapter has given a brief presentation of a new biosensor system for analysis of biomolecular interactions. The application examples shown were selected to illustrate some basic features of the technique. Many other applications, including receptor–ligand and protein–DNA interactions, the use of lectins as glycosylation probes and interactions between antibodies and intact virus particles, are in the process of being published.

Acknowledgements

Figures 7 to 10 were reproduced with kind permission from John Wiley & Sons, Ltd.

References

Andrade, J.D. (1986) *Surface and Interfacial Aspects of Biomedical Polymers: Surface Chemistry and Physics.* Plenum Press, New York.

Atassi, M. Z. (1984) *European J. Biochem.* **145**, 1–20.

Azimzadeh, A. and Van Regenmortel, M. H. V. (1990) *J. Mol. Recognition* **3**, 108

Bain, C. D., Troughton, E. B., Tao, Y.–T., Evall, J., Whiteside, G. M. and Nuzzo, R. G. (1989) *J. Am. Chem. Soc.* **111**, 321.

Benjamin, D. C., Berzofsky, J. A., East, I. J., Gurd, F. R. N., Hannum, C., Leach, S. J., Margoliash, E., Michael, J. G., Miller, A., Prager, E. M., Reichlin, M., Sercarz, E. E., Smith–Gill, S. J., Todd, P. E. and Willson, A. C. (1984) *Ann. Rev. Immunol.* **2**, 67–101.

Boardman, A. D. (1982) *Electromagnetic Surface Modes.* John Wiley and Sons, Chichester.

Burnens, A., Demotz, S., Corradin, G. Binz, H. and Bosshard, R. (1987) *Science* **235**, 780–783.

Crawford, G. D., Correa, L. and Salvaterra, P. M. (1982) *Proc. Natl. Acad. Sci. USA* **79**, 7031–7035.

Cullen D. C., Brown, R. and Lowe, C. (1987) *Biosensors* **3**, 211–225.

Dandliker, W. B. andLevison, S. A. (1967) *Immunochemistry* **5**, 171

Daniels, P. B., Deacon, J. K., Eddowes, M. J. and Pedley, D. G. (1988) *Sensors and Actuators* **15**, 11–18.

Fägerstam, L. G., Frostell, Å., Karlsson, R., Kullman, M., Larsson, A., Malmqvist, M. and Butt, H. (1990) *J. Mol. Recognition* **3**, 208–214.

Flanagan, M. T. and Pantell, R. H. (1984) Electron. Letters **20**, 968–970.

Froese, A. (1968) *Immunochemistry* **5**, 253.

Geysen, H. M., Mason, T. M. and Rodda, S. J. (1988) *J. Mol. Recognition* **1**, 32–41.

Hopp, T. P. and Woods, K. R. (1981) *Proc. Natl. Acad. Sci. USA* **78**, 3824–3828.

Johnstone, R. W., Andrew, S. M., Hogarth, M. P. Pietersz, G. A. and McKenzie, I. F. (1990) *Mol. Immunology* **27**, 327

Kooyman, R. P. H., Kolkman, H., Van Gent, J. and Greve, F. (1988) *Anal. Chim. Acta* **213**, 35–45.

Kronick, M. N. and Little, W. A. (1975) *J. Immunol. Methods* 8, 235–240.

Liedberg, B., Nylander, C. and Lundström, I. (1983) *Sensors and Actuators* **4**, 229–304.

Löfås, S and Johnsson, B. (1990) *J. Chem. Soc. Chem. Commun.* **21**, 1526–1528.

Mason, D.W. and Williams, A.F. (1986) In *Handbook of experimental Immunology. Vol 1. Immunochemistry.* (D. Weir, ed.). Blackwell Scientific Publications, Oxford.

Mazza, M. M. and Retegui, L. A. (1989) *Mol. Immunol.* **26**, 231–240.

Ngwainbi, J. (1990) *Biosensors and Bioelectronics* **5**, 13.

Nuzzo, R. G. and Allara, D. J. (1983) *J. Am. Chem. Soc.* **105**, 4481.

Olson, W. C., Spitznagel, T. M. and Yarmush, M. L. (1989) *Mol. Immunol.* **26**, 129

Raether, H. (1988) *Surface Plasmons on Smooth and Rough Surfaces and on Gratings.* Springer Verlag, Berlin.

Skubitz, K. M. and Smith, T. W. (1975) *J. Immunology* **114**, 1369

Stanley, C., Lew, A.M. and Steward, M.W. (1983) *J. Immunol. Methods* **64**, 119

Stenberg, E., Persson, B., Roos, H. and Urbaniczky, C. (1991) *J. Colloid Interface Sci.* **143**, 513–526.

Steward, M.W. (1986) *Handbook of Experimental Immunology. Immunochemistry.* Blackwell Scientific Publications, Oxford.

Van Regenmortel, M. H. V. and de Marcillac, G. D. (1988) *Immunol. Letters* **17**, 95–108.

Wilson, J. E. and Smith, A. D. (1984) *Anal. Biochem.* **143**, 179–187.

Yarmush, D. M., Morel, G. and Yarmush, M. L. (1987) *J. Biochem. Biophys. Methods* **14**, 279–289.

9

Application of Recombinant Antigens for the Diagnosis of Acute Epstein–Barr Virus Infection

Walter Hinderer and Günter Siegl

I. INTRODUCTION

A. Infectious mononucleosis

Primary infection with Epstein–Barr virus (EBV) can manifest as infectious mononucleosis (= mononucleosis infectiosa, morbus Pfeiffer, kissing disease, glandular fever) and can present with fever, pharyngitis, tonsillitis, lymphadenopathy, malaise, headache, myalgia, spleno- and hepatomegaly, rash, and leucocytosis. During childhood an EBV primary infection is mostly a mild or asymptomatic event, but even in adults severe complications are rarely observed. In the course of primary infection B–lymphocytes as well as oropharyngeal and probably also nasopharyngeal and salivary epithelial cells will be infected, and virus may persist in such cells for the entire life of the individual. Approximately 90% of the human population is infected and is thus capable of transmitting the virus for life. In general, the transmission does not occur by flighty contacts, but needs a close connexion of oral surfaces, e. g. during kissing (for reviews see Cheeseman, 1988 and Sumaya, 1989).

B. EBV diagnosis: The classical antigen complexes

Laboratory diagnosis of EBV infection is accomplished either by detection of viral DNA in tissue material using *in situ* nucleic acid hybridization techniques (zur Hausen *et al.*, 1970) or serologically, i.e. by detection of EBV–specific antibodies. Diagnostic serology makes use of the fact, that several distinct virus–

Walter Hinderer • BIOTEST AG, Research Dept., Landsteinerstr. 5, D–63033 Dreieich, Germany *Günter Siegl* • Institute for Clinical Microbiology and Immunolology, Frohbergstr. 3, CH–9000 St. Gallen, Switzerland.

Applied Virology Research, Vol. 3, Edited by
E. Kurstak *et al.*, Plenum Press, New York, 1994

specific antigen complexes are expressed at various stages during the EBV replication cycle. Antibodies directed against these antigens are also characteristic for different stages of EBV infection and, hence, can be used to distinguish between acute, reactivated or past infection. The EBV antigen complexes have been defined by immunofluorescent staining and the respective antibodies are detected in serum samples by indirect immunofluorescence assays (Reedman and Klein, 1973, Henle *et al.*, 1974). The basis of these assays are fixed cells derived from various cell cultures infected by distinct strains of EBV and expressing constitutively (phase of latency) or after induction with chemicals, immunoglobulins, or superinfection, the antigens in a time–dependant manner (early lytic phase, late lytic phase). The diagnostically used antigen complexes are designated Epstein–Barr nuclear antigens (EBNA), early antigens (EA), virus capsid antigens (VCA), and membrane antigens (MA). Each class is composed of many distinct EBV encoded polypeptides and except the group of EBNAs these antigens are, so far, not fully defined at the protein level (for a review see Dillner and Kallin, 1988). The EAs can be subdivided further into EA–D (diffuse) and EA–R (restricted), due to their cellular distribution following methanol fixation (Henle *et al.*, 1971).

Up to now, the immunofluorescent (IF) methods are the preferred assays for specific EBV diagnosis and frequently are used to obtain serological evidence for acute EBV infection. However, these assays are time consuming, labor intensive, require educated personal and are not suitable for large–scale testing and automatic handling. In addition, standardization is difficult, due to the variability of the antigen producing cell cultures as well as to subjective reading of results. Furthermore, the interference of autoantibodies is a common problem of assays using non–purified antigenic substrates.

C. ELISA technology

Various attempts have been made to simplify and to improve EBV serodiagnosis by developing microtest plate ELISAs for detection of specific antibodies. Until recently, the overwhelming majority of such ELISAs were based on the cell culture derived viral antigens, by use of either partially purified antigen complexes or, with the help of monoclonal antibodies, affinity–purified viral proteins (Pearson, 1988). However, one problem of these assays is related to the difficulty in producing the antigens for large–scale use (Pearson, 1988).

Alternatively, synthetic peptides may constitute a reliable source of antigen (Fox *et al.*,1987, Geltofsky *et al.*,1987). They can be selected, for example, by epitope–mapping of the respective protein (Middeldorp and Meloen, 1988). In general, however, only highly immunogenic linear epitopes can be used for this purpose. For EBV such immunodominant sequences are the glycine–alanine copolymer of ENBA–1, the polyproline part of EBNA–2 (Linde *et al.*, 1990), and the C–terminal region of VCA–p18 (Van Grunsven *et al.*,1993a).

Finally, EBV defined EBV polypeptides can be produced in sufficient quantities and quality by use of recombinant DNA technology. With this technique it is also possible to gain increased specificity by selecting and combining diagnostically valuable proteins on the one hand, and omitting putative cross reacting regions on the other hand. The test system described below has been developed along this line.

D. Recombinant EBV antigens

Using immunoprecipitation and Western blot experiments Wolf and coworkers (Wolf *et al.*,1985, Motz *et al.*,1986) characterized several EBV–associated antigens. Of these, four polypeptides, EA–D–p54, EA–p138, VCA–p150, and EBNA–1–p72, were selected for further analysis. The coding sequences were cloned and expressed in E. coli without any fused bacterial protein moieties and the polypeptides were purified near homogeneity (Hinderer *et al.*, 1988). The products thus obtained have the following characteristics (see Table 1): The p54–clone contains the complete coding sequence. Computer–assisted predictions of putative immunogenic regions led to the construction of truncated autologous fusion proteins in the case of the large antigens p150 and p138 (Motz *et al.*,1986). The p72–clone contains the coding region of the carboxy–half of EBNA–1 expressing 46.2% of the amino acids without the already mentioned, glycine–alanine copolymer. The latter structural feature of EBNA–1 is known to cross–react with autoantibodies, most propably due to sequence similarities with collagen and keratin (Baboonian *et al.*, 1989, Birkenfeld *et al.*, 1990). In all instances high–grade purification is absolutely necessary, because antibodies directed against components of enterobacteria are widespread in human sera. Likewise, fusion proteins containing bacterial residues, e.g. β–galactosidase, which are often used to obtain a stable expression, are not recommended for use in diagnostic applications.

Table 1. Recombinant EBV antigens

Protein	p54	p138	p150	p72
Reading frame	BMRF 1	BALF 2	BcLF 1	BKRF 1
Classification	Early antigen (EA–D)	Early antigen ("major DNA–binding protein")	Virus capsid antigen (VCA)	Nuclear antigen (EBNA–1)
Construction	Whole coding region	Fusion of two selected sections	Fusion of two selected sections	C–terminal part (46.2%)
Size of expressed protein (Mr)	52/47 kD	39 kD	47/22 kD	46 kD
Concentration in coating solution	1 µg/ml	1 µg/ml	10 µg/ml	0.25 µg/ml

Studies (Hinderer *et al.*, 1988) have demonstrated an excellent diagnostic potential of the recombinant antigens; yet, they also indicated, that the different antigens had to be coated in varying concentrations to compensate for differences in reactivity. In a more recent, more detailed investigation (Gorgievski–Hrisoho *et al.*, 1990) we evaluated the four recombinant EBV proteins for their potential to detect and to measure specific IgM and IgG responses in cases of well–defined infectious mononucleosis with special focus on the early acute phase of the disease. The diagnostic significance of these recombinant antigens and the implications of the results of the latter study for the design of a new generation EBV ELISA system is presented in the following.

II. EXPERIMENTAL

A. Antigens

The selected genomic regions were cloned and directly expressed in *E. coli* JM109 using standard vectors pUC. All four clones expressed stable proteins (Table I). The recombinant antigens were produced at a preparative scale as described previously (Hinderer *et al.*, 1988, Gorgievski–Hrisoho *et al.*, 1990). Purification was monitored and quality control was performed by SDS–polyacrylamide gel electrophoresis and by Western blotting using defined human sera, antigen–specific and anti–*E. coli* rabbit sera, and monoclonal antibodies. The purity of the antigens used for coating was near homogeneity, i.e. *E. coli* contaminants are far below 1%.

B. Sera

Serum samples were collected from patients, who presented with clinical symptoms suggestive for infectious mononucleosis. Symptoms commonly found in these patients included fever, tonsillar enlargement, lymphadenopathy, spleno/hepatomegaly, hepatitis, rash, and atypical lymphocytes. The majority of the sera were heterophile antibody negative. Those cases, which had been identified by a positive VCA–IgM result (>1:32) were included in the figures and tables (33 patients, aged 2–32 y., mean: 16.2 y.). From 13 of these patients 2–4 follow–up samples were available. In addition, sera of 161 asymptomatic, apparently healthy blood donors formed a control panel.

C. Standard laborararatory diagnosis

IgG antibodies to VCA and EA were determined by indirect immunofluorescence (IF), according to Henle *et al.*, 1974, on the basis of slides prepared from antigen producing P3HR1 cells or by use of slides purchased from Gull Laboratories, USA. IgM antibodies to VCA were assayed by indirect immunoperoxidase (IP) staining on slides obtained from Savyon Diagnostics, Israel. To remove non–specific reactivities, due to the presence of rheumatoid factors and antigen–specific IgG (Henle *et al.*, 1979), the sera were preadsorbed as specified by the manufacturer of the test system. Antibodies against EBNA were determined by anticomplement enhanced immunofluorescence (ACIF), according to Reedman and Klein (1973) on slides obtained from Gull Laboratories.

D. ELISA methods

The principle of the indirect ELISA was used. The purified antigens were immobilized on the solid phase (polystyrene microtest plates, 96 wells, Nunc, Denmark) in concentrations as specified in Table 1. Identical plates were used for IgM and IgG testing. Murine monoclonal antibodies, anti–human IgG (Biotest AG, FRG) and anti–human IgM (Janssen, Belgium), both conjugated with horse radish peroxidase, served as second antibodies. The incubations were as follows: Serum (1:21) for 30 min (IgG) or 60 min (IgM) at 40 °C; second antibody for 30 min at 40 °C, substrates (1,2 phenylenediamine and H_2O_2) for 15 min at room temperature. All volumes were 100 µl. The absorbance values were measured at 492/620 nm. The sera were tested in duplicate and a blank, positive and negative control were included on each plate. To avoid false–positive results in IgM determinations due to the presence of rheumatoid factors (Henle *et al.*, 1979), sera were treated over night at 2–8 °C with a polyclonal anti–human IgG–Fc goat serum. The cut–off values were defined as OD of the negative control + 0.15.

III. RESULTS

The four individual EBV antigens were evaluated with respect to IgM and IgG reactivity in direct comparison with standard IF/IP assays. Altogether, eight different ELISAs were tested with two panels of sera, which were derived from 33 IM patients and 161 blood donors.

A. Antibody reactivity in sera of blood donors

The right hand panels of Figures 1 and 2 show the mean OD values and the percentage of positives, respectively, in sera of asymptomatic blood donors. Significant positive ELISA reactions with antigens p54, p138 and p150 were further investigated by Western blot experiments and, for the most part, also by IF/IP testing. IgM reactivities in these sera were rare, weak, and could not be confirmed in Western blots. IgG antibodies were present at higher prevalence with anti–p72 (EBNA–1)–IgG being present at highest frequency and highest mean OD value. These antibodies were detectable in 86% of the blood donors. About half of those donors which lack p72–IgG, were proven by further assays to be EBV seronegative, i.e., are most likely non–infected individuals. Significant IgG–reactivity with p54, p138, and p150 has been observed in some few sera only. Most of these specimens had weak OD values.

Altogether, sera of healthy individuals with an EBV infection in the past characteristically lack specific IGM antibodies against the recombinant antigens under test. Occasionally, they contain low IgG titers against p54, p138 and p150, but almost regularly show anti–p72–IgG in high concentration (Gorgievski–Hrisoho *et al.*, 1990).

B. Antibody reactivity in sera of IM patients

As shown in the left hand panels of Figures 1 and 2, IgM antibodies binding to p138 and p54 were detectable in IM sera with high frequency and high OD values. These antibodies thus appear to be the most discriminating ones for acute and past EBV infection. Anti–p150–IgM was present at a lower prevalence and with moderate OD values only. The sensitivity of the p72–IGM ELISA was too weak. With a rather low antigen concentration this assay has been adapted to preferentially detect IgG antibodies in accordance with the well–known delay in EBNA–1–IgG seroconversion observed in IF assays (Henle *et al.*, 1987).

Anti–p138–IgG and anti–p54–IgG coincided with IgM antibodies. Both markers were highly prevalent in IM sera (Fig. 2). The absorption values were usually high. Significant differences in individual patients have also been observed as, for instance, some patients were positive either for anti–p138 or anti–p54–IgG. A certain qualitative but no direct quantitative correlation became evident between EA–IF titers and absorption values of the p54– or p138–IgG–ELISA. Compared to EBNA–ACIF the p72 (EBNA–1)–IgG–ELISA revealed an increased sensitivity. Although the reference assay used (ACIF) is not specific for EBNA–1 and IgG, the results of the p72–IgG–ELISA correlated well with those of the EBNA–ACIF tests. The p72–IgG–ELISA typically yielded negative results during primary acute EBV infections (fig. 2). The p150 (VCA)–IgG–ELISA proved to be of surprisingly low sensitivity, and more importantly, failed to discriminate between recent and past infection (Fig. 1 and 2).

In summary, acute EBV infections are characterized by the presence of anti–p138–IgM and anti–p54–IgM, the presence of anti–p138–IgG and/or p54–IgG, and a lack of anti–p72–IgG antibodies (Gorgievski–Hrisoho *et al.*, 1990).

C. Seroconversions

Some of the follow–up panels displayed seroconversions when tested by VCA–IgM–IP or EBNA–ACIF. Such sera could be used to gain information on the sensitivity of the recombinant ELISAs. In three out of five seroconversions detected by the VCA–IgM–IP assay (Table 2), p54–IgM and/or p138–IgM were

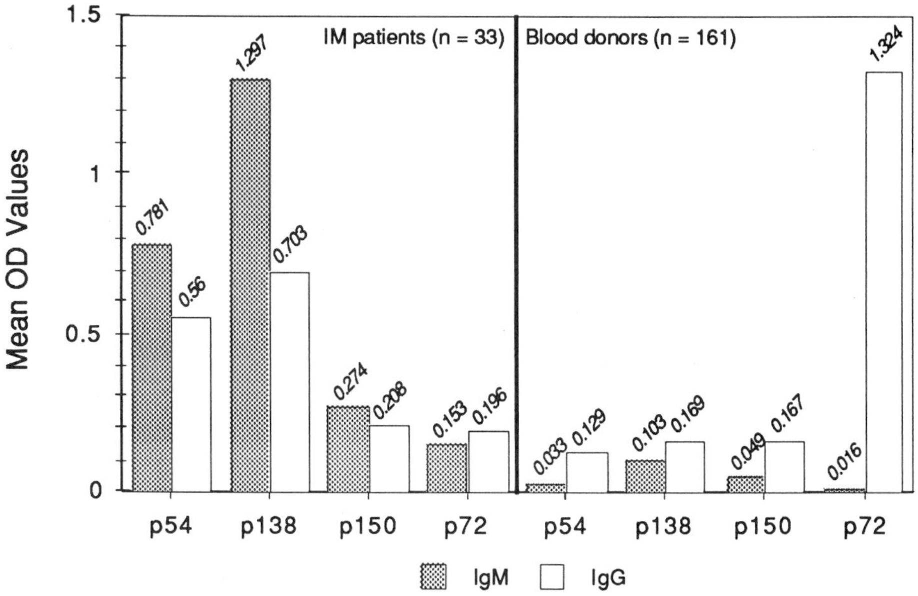

Figure 1.
Mean OD values of the recombinant ELISAs.

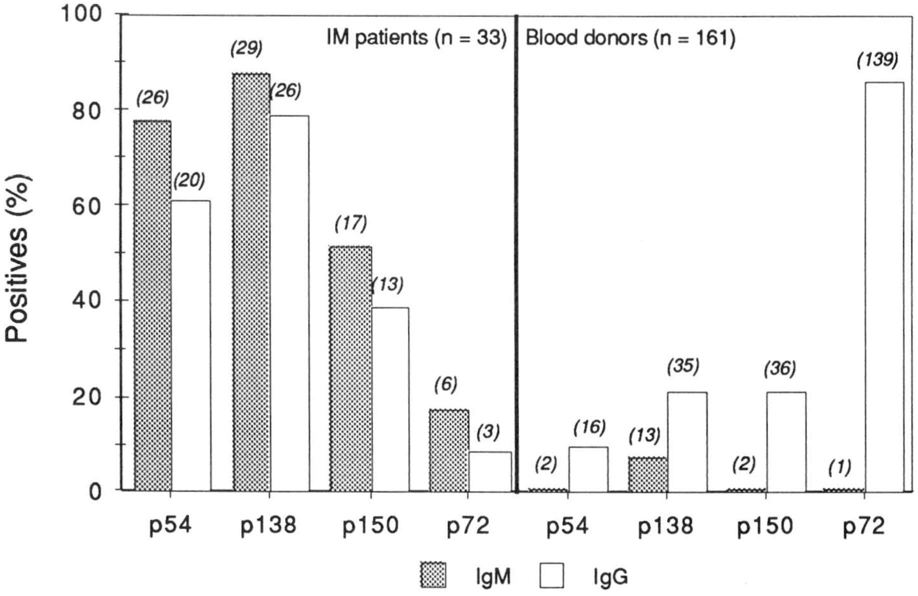

Figure 2.
Positive rates (%) of the recombinant ELISAs.

Table 2. ELISA Results of VCA–IgG–IPA Seroconversions

Age of Patient	Days[b]	IPA[a] VCA[c] IgM	ELISA (OD–values) p54 IgM	p54 IgG	p138 IgM	p138 IgG	p150 IgM	p150 IgG	p72 IgM	p72 IgG
26	0	–	0.518	0.058	0.275	0.370	0.058	0.394	0.0210	0.042
	8	+	2.118	1.406	2.475	1.208	0.240	0.460	0.036	0.049
22	0	+/–	0.049	0.028	0.086	0.153	0.069	0.144	0.085	0.114
	8	+	0.783	0.361	0.884	0.714	0.111	0.118	0.009	0.009
13	0	+/–	0.212	0.040	0.947	0.533	0.194	0.148	0.296	0.023
	6	+	0.625	0.059	1.008	1.085	0.178	0.193	0.361	0.031
	24	+/–	1.836	0.394	1.676	1.837	0.252	0.155	0.131	0.023
12	0	+/–	0.167	2.889	2.387	2.488	0.110	0.269	0.214	0.004
	254	+	0.200	2.645	1.445	2.058	0.122	0.213	0.072	0.009
	264	+	0.116	2.442	0.737	0.767	0.095	0.200	0.144	0.005
4	0	–	0.005	0.008	0.053	0.038	0.003	0.149	0.008	0.003
	1	+	0.029	0.028	0.387	0.109	0.045	0.404	0.054	0.013
	10	+	0.148	0.340	0.145	0.815	0.105	0.360	0.040	0.164
	23	+	0.131	0.458	0.074	0.548	0.088	0.227	0.040	0.086

a) immunoperoxidase assay; b) days after first sample; c) + = >1:32, +/– = 1:32, – = <1:32

Table 3. ELISA Results of EBNA–ACIF Seroconversions

Age of Patient	Days[b]	ACIF[a] EBNA[c]	ELISA (OD–values) p54 IgM	p54 IgG	p138 IgM	p138 IgG	p150 IgM	p150 IgG	p72 IgM	p72 IgG
25	0	–	0.452	0.380	0.805	1.032	0.189	0.273	0.091	0.000
	112	+/–	0.102	0.051	0.179	0.129	0.071	0.191	0.051	0.249
	283	+	0.050	0.032	0.193	0.085	0.081	0.160	0.049	2.179
15	0	–	0.047	0.678	0.078	0.107	0.052	0.064	0.039	1.740
	18	+/–	0.382	0.300	0.144	0.100	0.175	0.048	0.143	0.1633
6	0	–	0.265	0.032	0.353	0.214	0.507	0.132	0.177	2.588
	48	+	0.208	0.024	0.234	0.189	0.388	0.116	0.127	2.617

a) antik–complement immunofluorescence; b) days after first sample; c) + = >1:10, +/– = 1:10, – = <1:10

already clearly positive in the initial serum specimen, which still yielded negative or equivocal results in VCA–IgM–IP assay. Positive results with the remaining two panels were obtained simultaneously in the second sample with both test systems. These findings point to an increased sensitivity of the recombinant IgM–ELISA, a notion supported further by ELISA–identification of three additional EBV primary infections among 23 patients with typical IM symptoms, but negative in VCA–IgM. Following this identification, extended testing of the latter sera with traditional methods yielded negative EBNA–ACIF and positive EA– and VCA–IgG–IF results (Gorgievski–Hrisoho *et al.*, 1990).

Two of the three EBNA–ACIF seroconversion panels presented in Table 3 were initial positive with the p72–IgG–ELISA. This assay is more sensitive than ACIF and highly specific for IgG antibodies directed against the carboxy–half of EBNA–1. Using plates coated with increased antigen concentration or Western blot experiments, a much higher proportion of IM sera, even of those collected in a very early phase of infection, were found to be p72–IgG positive. It might, therefore, be speculated that the so far accepted delay in EBNA–1–IgG seroconversion (Henle *et al.*, 1987) reflects the insensitivity of the IF method rather than a true absence of these antibodies early in infection. The value of a super sensitive p72–IgG ELISA for the diagnosis of primary infections is disputable, however, for the definition of seropositivity it may have advantages.

IV. DISCUSSION AND CONCLUSIONS

Our current knowledge concerning the serodiagnostic potential of the recombinant EBV proteins is summarized in Table 4. According to their significance in various phases of an EBV infection, the specific antibodies can be devided into three groups: i) IgM antibodies against all recombinant proteins can be detected early during primary infection and disappear rapidly during convalescence. ii) IgG antibodies against p54, p138 and p150 appear immediately after IgM antibodies, but they decline more slowly and can persist in some individuals indefinitely for a long time. iii) Anti–p72–IgG shows a marked delay in appearance, reaches highest values after convalescence, and most likely remains positive for life. Therefore, anti–p72–IgG like VCA–IgG can be taken as an indicator of past (and continuing latent) EBV infection.

As already mentioned Western and dot blot experiments revealed that nearly all serum specimens from IM patients contained anti–p72–IgM. This, however, was not detected in the described ELISA due to the low antigen concentration applied (Table 1). We do not intend to adjust test conditions in a way that anti–p72–IgM can also be determined by ELISA. In such a test difficulties with false–positive reactions are likely, because the usually very high anti–p72–IgG titers in individuals with past EBV infection would support the interference by rheum–atoid factors (Henle *et al.*, 1979). Although such reactivities can be omitted by pretreatment of the sera with IgG blocking reagents or by changing of the test format (μ–capture principle), this is not necessary as IgM antibodies against p54 and p138 alone allow a significant diagnostic decision.

Table 4. Prevalence of antibodies (diagnostic scheme)

Stage of Infection	p54, p138, p150, (p72) IgM	p54, p138, (p150) IgG	p72 IgG	Comments
Susceptible	–	–	–	Also seronegative during early phase of incubation period
Acute primary infection	+	+ or –	–	Typically strong reactions
Acute primary infection or reactivation	+	+	+	Acute primary infection: At most weak EBNA
Post–acute infection (convalescence)	–	+	–	Chronic infections may persist with this pattern
Past (latent) infection	–	– or +	+	Strong EA IgG occurs rarely

The reactivity of the VCA–p150 substrate was not as good as expected. Despite the high antigen concentration applied (Table 1), specific antibodies were detectable with relatively low frequencies only. Anti–p150–IgM showed a certain diagnostic potential, but anti–p150–IgG evidently failed to contribute substantially to the diagnosis of EBV infections. Consequently this antigen has not been considered further. In this context it is of interest to note that native p150 (BcLF1) from virus–producing cells likewise demonstrated restricted IgM reactivity (Pearson and Luca, 1986). The p150 is the major viral capsid component, however, the dominant antigen of the VCA complex in current assay systems was shown to be the envelope protein gp125 (BALF4), a homologue to herpes simplex virus glycoprotein B (HSV–gB) and varicella zoster virus glycoprotein II (VZV–gpII); Pearson and Luka, 1986). Recently, two additional capsid proteins, VCA–p40 (BdRF1) and VCA–p18 (BFRF3), have been identified (Van Grunsven *et al.*, 1993a). These antigens demonstrated better IgG reactivity than p150 in Western blot studies among sera from seropositive donors (Van

Grunsven *et al.*, 1993b). It remains to be seen, if p40 and p18 are diagnostically useful representatives of the serologically defined VCA complex.

Finally, two important facts should be emphasized. Firstly, to provide against the observed differences in individual patients a combination of antigens is required. Secondly, the use of more than one diagnostic assay is recommended to gain a suggestive picture of the EBV immune status at the various stages of infection. On the basis of results obtained in these and further studies (Hinderer *et al.*, 1991) we decided to develop three different ELISAs: A mixture of p54 and p138, representative for EBV early antigens, is used to measure both specific IgM and IgG antibodies in two separate tests (EA–IgM and EA–IgG ELISA). In addition, p72 forms the basis of a second IgG assay (EBNA–IgG ELISA). This novel EBV ELISA system, designated "BIOTEST anti–EBV recombinant", has been tested recently in a clinical study in direct comparison with classical immunofluorescence assays (Faerber *et al.*, 1993), and included 120 patients with infectious mononucleosis, 60 patients with acute HCMV infection, toxoplasmosis or rheumatic disease, and 185 healthy donors. The results suggest that the recombinant ELISA system can in fact be used for a standardized, rapid, and highly sensitive diagnosis of acute primary EBV infection. Advanced studies will also concentrate on further EBV–associated diseases, i.e. chronic infection, reactivation in immunocompromised patients (transplantation, HIV) and nasopharyngeal carcinoma. In conclusion, the recombinant ELISA system proved to be an alternative method to standard immunofluorescence serology, but in addition offers advantages, e.g. simple and unique performance, objective reading of results, reduced interference with rheumatoid factors, and no interference with anti–cellular antibodies (Faerber *et al.*, 1993).

REFERENCES

Baboonian, C., Halliday, D., Venables, P. J. W., Pawlowski, T., Millman, G. and Maini, R. N. (1989). *Rheumatol. Intl.* **9**, 161–166.

Birkenfeld, P., Haratz, N., Klein, G. and Sulitzeanu, D. (1990). *Clin. Immunol. Immunopathol.* **54**, 14–25.

Cheeseman, S. H. (1988). *Sem. Hematol.* 25, 261–268.

Dillner, J. and Kallin, B. (1988). In *Advances in Cancer Research Vol. 50* (G. Klein and S. Weinhouse, eds.), pp. 95–158, Academic Press, San Diego.

Faerber, I., Wutzler, P., Wohlrabe, P., Wolf, H., Hinderer, W. and Sonneborn, H.–H. (1993). *J. Virol. Method.* **42**, 301–308.

Fox, R. I., Scott, S., Houghten, R., Whalley, A., Geltofsky, J., Vaughan, J. and Smith, R. (1987). *J. Clin. Lab. Anal.* **1**, 140–145.

Geltofsky, J. E., Smith, R. S., Whalley, A. and Rhodes, G. (1987). *J. Clin. Lab. Anal.* **1**, 153–162.

Gorgievski–Hrisoho, M., Hinderer, W., Nebel–Schickel, H., Horn, J., Vornhagen, R., Sonneborn, H.–H., Wolf, H. and Siegl, G. (1990). *J. Clin. Microbiol.* **28**, 2305–2311.

Henle, G., Henle, W. and Klein, G. (1971). *Intl. J. Cancer* **8,** 272–282.

Henle, G., Lenette, E. T., Alspaugh, M. A. and Henle, W. (1979). tl **36,** 415–422.

Henle, W., Henle, G. andersson, J., Ernberg, J., Klein, G., Horwitz, C. A., Marklund, G., Rymo, L., Wellinder, C. and Straus, S. E. (1987). *Proc. Natl. Acad. Sci. USA* **84,** 570–574.

Henle, W., Henle, G. and Horwitz, C. A. (1974). *Hum. Pathol.* **5,** 551–556.

Hinderer, W., Nebel-Schickel, H., Sonneborn, H.-H., Motz, M., Kühbeck, R. and Wolf, H. (1988). *J. Exptl. Cancer Res.* **7,** 132.

Hinderer, W., Nebel-Schickel, H., Horn, J., Vornhagen, R., Sonneborn, H.-H., Gorgievski-Hrisoho, M., Siegl, G., Faerber, I., Wutzler, P. and Wolf, H. (1991). In *Epstein-Barr Virus and Human Disease 1990* (D. V. Ablashi, A. T. Huang, J. S. Pagano, G. R. Pearson and C. S. Yang, eds.), pp. 123–125, Humana Press, Clifton.

Linde, A., Kallin, B., Dillner, J. Andersson, J., Jägdahl, L., Lindvall, A. and Wahren, B. (1990). *J. Infect. Dis.* **161,** 903–909.

Middeldorp, J. M. and Meloen, R. H. (1988). *J. Virol. Method.* **21,** 147–159.

Motz, M., Fan, J., Seibl, R., Jilg, W. and Wolf, H. (1986). *Gene* **42,** 303–312.

Pearson, G. R. (1988). *J. Virol. Method.* **21,** 97–104.

Pearson, G. R. and Luka, J. (1986). In *The Epstein-Barr Virus: Recent Advances* (M. A. Epstein and B. G. Achong, eds.), pp. 48–74, W. Heinemann, London.

Reedman, B. M. and Klein, G. (1973). *Intl. J. Cancer* **11,** 499–520.

Sumaya, C. V. (1989). *J. Clin. Immunoassay* **12,** 168–174.

Van Grunsven, W. M. J., Van Heerde, E. C., De Haard, H. J. W.,Spaan, W. J. M. and Middeldorp, J. M. (1993a). *J. Virol.* **67,** 3908–3916.

Van Grunsven, W. M. J., Nabbe, A. and Middeldorp, J. M. (1993b). *J. Med. Virol.* **40,** 161–169.

Wolf, H., Motz, M., Kühbeck, R., Seibl, R., Bayliss, G. J., Modrow, S. and Fan, J. (1985). In *Developements in Medical Virology Vol. 1: Epstein-Barr Virus and Associated Diseases* (P. H. Levine, D. V. Ablashi and G. R. Pearson, eds.), pp. 485–494, M. Nijhoff Publishing, Boston.

zur Hausen, H., Schulte-Holthausen, H., Klein, G., Henle, W., Henle, G., Clifford, P. and Santesson, L. (1970). *Nature* **288,** 1056–1058.

Automation of the Detection of DNA Sequences for the Laboratory Diagnosis of Viral Infections

L. Wray

I. INTRODUCTION

The ability to directly detect nucleic acid sequences from infected biological fluids and tissue offers significant advantages in the clinical management of viral disease. In particular, a more precise and timely diagnosis is rendered possible, with the resultant advantages in treatment. For the clinical laboratory, time and labor can be reduced, procedures simplified, and organisms which are fastidious to culture can be identified more readily (Tenover, 1988).

The introduction of DNA probe tests into routine clinical laboratory procedures has been hindered for a number of reasons (Landegren, *et. al.*, 1988, Matthews and Drika, 1988). The sensitivity of these procedures has been inadequate compared with conventional culture techniques or immunoassays. To maximize sensitivity radioactive labels have been used, with the accompanying problems of laboratory management of radioisotopes and their disposal. To some degree, non–radioisotopic methods have been able to serve as alternatives, but have either not had adequate sensitivity or have been cumbersome or laborious to run. To achieve maximum sensitivity, the incubation periods using current methodology have been long, taking several hours or days. The tests have also employed multiple time consuming steps, requiring in many cases considerable previous training and expertise. Automation is clearly a prerequisite for the full benefits of this technology to be realized in the clinical laboratory.

The ability to amplify DNA from clinical samples has allowed diagnostic tests to be developed with DNA probe technology which provide the sensitivity required for clinical utility in a nonisotopic format. In fact, the sensitivity now afforded by probe tests is greater than for any other method and approaches the maximum theoretically possible, a single molecule of DNA, one viron! A number of methods have been developed for amplifying DNA, and have been extensively reviewed previously (Kwoh et al., 1990; Anonymous, 1991).

L. Wray • Probe Diagnostics, Abbott Laboratories, Abbott Park, Illinois, USA 60064–3500

Applied Virology Research, Vol. 3, Edited by
E. Kurstak *et al.*, Plenum Press, New York, 1994

II. POLYMERASE CHAIN REACTION

PCR (Saiki *et. al.*, 1985) (Fig. 1) employs two primers which anneal to complementary strands of DNA which has been thermally denatured. The primers are then extended with DNA polymerase (and nucleotide triphosphates), yielding two double stranded forms of DNA for the region bracketed by the primers. These can be denatured by heating and the process of primer annealing and extension repeated, again doubling the number of identical sequences. The process of heating to denature the DNA and cooling to copy the target sequence can be repeated over and over, with an amplification of one million fold being realized at about 25 cycles. The use of a thermostable enzyme from *T. aquaticus* (*Taq*), allows the process to be performed continuously without the addition of this reagent at each cycle.

III. LIGASE CHAIN REACTION

LCR (Bond *et. al.*, 1990, Barany, 1991), first described by Biotechnica International, Inc. in European Patent Publication EP–A–0320308, (Fig. 2) also employs thermocycling to amplify DNA. In this case however, four DNA probes are employed. These are complementary to the target sequence itself, rather than bracketing it as with the primers for PCR. After denaturation of the target DNA, if the complementary sequence is present, the probes hybridize, two on each strand, adjacent to one another, and complementary to the two probes which hybridize to the other strand. After hybridization, the two probes on each strand are joined with the enzyme, DNA ligase. This process can be repeated, as with PCR, with the ligated probes now serving as a template for a new round of hybridization and ligation. Comparable amplification to PCR can be achieved after the appropriate number of cycles. Exquisite specificity can be achieved with LCR due to the requirement of precisely adjacent binding of two probes on each strand for ligation to occur and subsequent amplification. A thermostable ligase from *T. thermophilus* is employed for uninterrupted cycling.

Immunoassay technology (Gosling, 1990) has progressed rapidly during the last two decades to the point of being essentially fully automated. Because the system which will be described below for automating probe tests employs this technology, some effort will be spent here outlining basic concepts and benefits. In particular, the so called two site or "sandwich" immunoassay will be described. This system employs two antibodies, directed at two separate sites on an analyte (Fig. 3), one antibody being immobilized to a solid phase (plastic beads, polystrene microparticles, microtiter plates), and the other labeled with a signal generator, which could be a radioisotope (P^{32}, I^{125}), a chemiluminescent compound such as acridine, or an enzyme (peroxidase, alkaline phosphatase).

The sample is first incubated with the solid phase, and any target for which the antibody has specificity is bound. The complex is washed, and then incubated with the second antibody containing the signal generator. Binding of

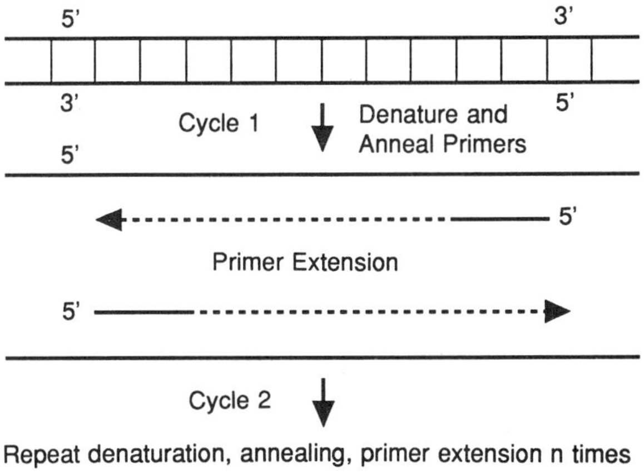

Figure 1.
Amplification of DNA by the polymerase chain reaction.

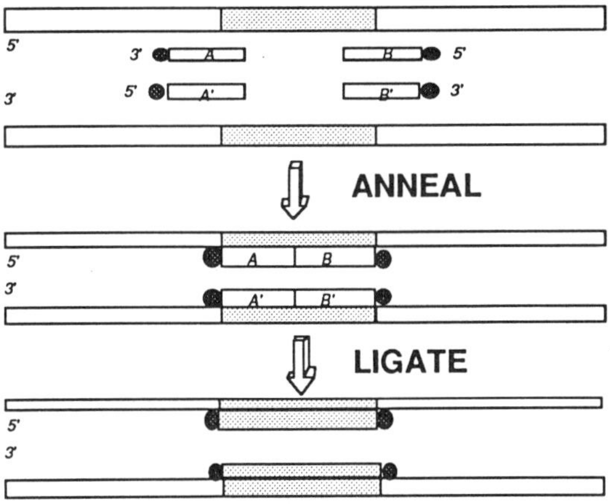

Figure 2.
Amplification of DNA by the ligase chain reaction (LCR)

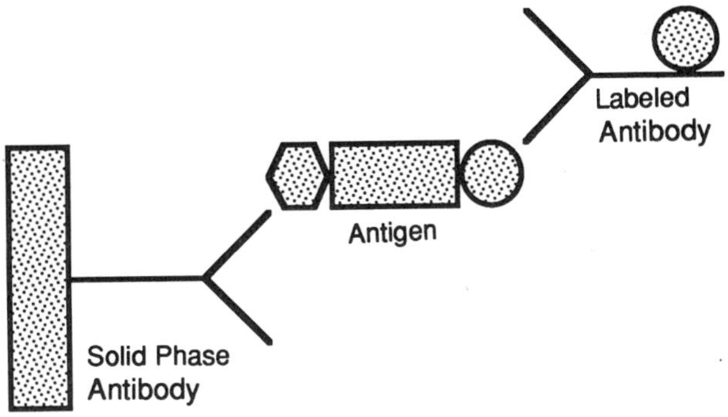

Figure 3.
General scheme for an immunoassay

Figure 4.
LCR product detection by immunoassay. A and B represent either fluorescein or biotin labels. AP is alkaline phosphatase.

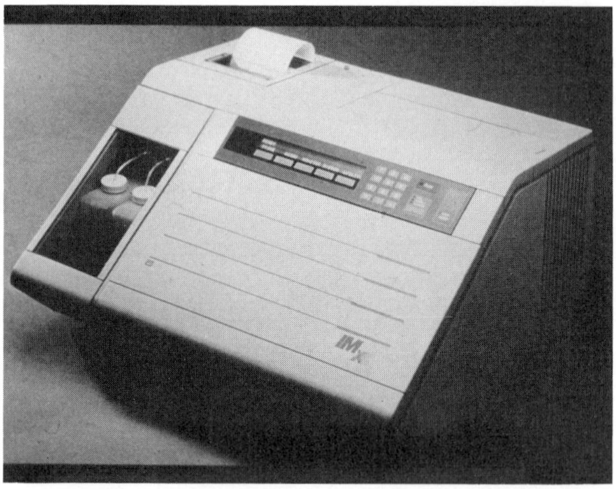

Figure 5.
The IMx automated immunoassay analyzer.

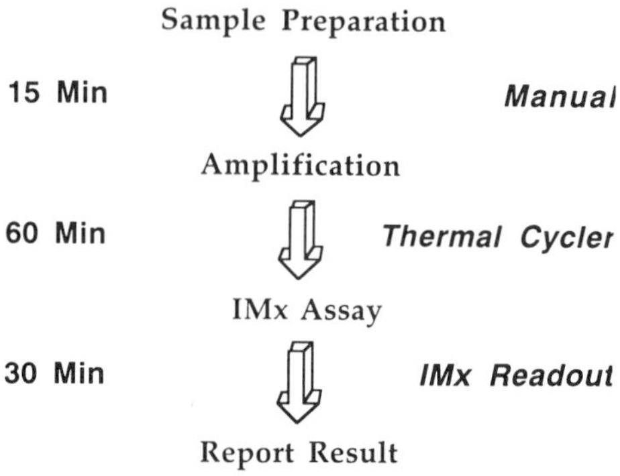

Figure 6.
Steps involved in performing semi–automated assays by LCR. Assay results can be obtained in less than two hours.

this second antibody can be detected by the presence of a colored reaction or fluorescence in the case of an enzyme label, or by radioactive emission or light for a radioisotope or chemiluminescent label respectively. Very high specificity is imparted by this two site format by the absolute requirement for the binding of both antibodies to the target being detected, thus minimizing any crossreactivity imparted by the individual antibodies. As mentioned above, all reagent addition, washing, and reading steps have been automated for a number of systems, to be discussed in greater detail below.

The approach to be described below adapts LCR amplification technology to an immunoassay format for semi–automated DNA probe assays.

To detect the presence of LCR product by immunoassay methods, the probes are labeled with biotin and fluorescein (Fig. 4). Upon ligation a double stranded DNA product is formed which then contains biotin at one end and fluorescein at the other and can serve as an analyte in a two site assay. In this particular example, the solid phase, consisting of microparticles coated with anti–fluorescein antibodies "captures' ligated product; after a wash step, a second biotin specific antibody, labeled with alkaline phosphatase binds to the biotin, which will yield a fluorescent signal when incubated with a suitable substrate such as 4–methylumbelliferyl phosphate (MUP). The signal generated can then be read with a fluorometer.

The above immunoassay procedure can be fully automated by employing the Abbott IMx^R system (Fiore *et al.*, 1988) (Fig. 5). This is a benchtop instrument for running assays in either a microparticle capture enzyme immunoassay (MEIA) format for macromolecules or fluorescence polarization immunoassay (FPIA) for hapten assays. The MEIA format is utilized for probe assays, with antibodies specific for haptens used to label the probes, as discussed above. The instrument is microprocessor controlled and all assay procedures, including pipetting and reagent additions, are performed automatically using a robotic arm. The sample is incubated with the microparticles in solution, which are then separated from the reaction mixture and washed on a glass–fiber matrix. The alkaline phosphatase labeled second antibody is then added to the matrix, and after washing and substrate (MUP) addition the signal is read with a front surface fluorometer. The processor uses linear regression analysis to convert the fluorescent measurements to rates, which are proportional to analyte concentration. The instrument can run up to 24 samples in a batch.

Cellular material is spun in phosphate buffered saline (PBS). The pellets are then resuspended in 100 µl of 10mM NaOH and heated to 100°C for 5 to 10 minutes. After cellular debris is removed by recentrifugation, 5 µl of the supernatant is removed and added to the other components of the reaction mix.

The reaction mix is prepared in siliconized, 500 µl microfuge tubes and contains, in addition to the sample: 50 mM EPPS (N–(2–hybroxy-ethyl)piperazine–N–(3 propanesulfonic acid), 1 mM DTT, 10 µg/ml BSA, 0.1mM NAD, 7.5 x 10^{11} molecules of probes A' and B, 5.0 x 10^{11} molecules of probes A and B' (Fig. 4). A and A' are derivatized with capture ligand, fluorescein, and B and B' with the signal moiety, biotin (Morgan and Celebuski, 1991). A' and B are

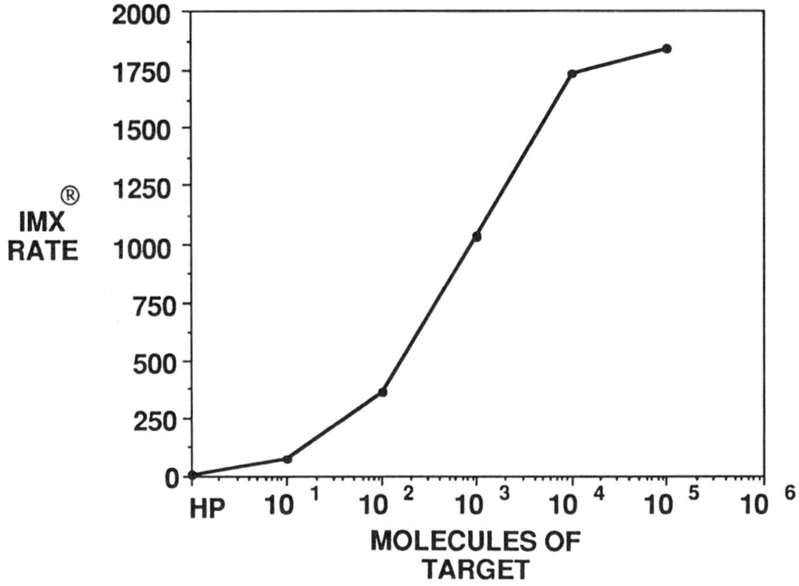

Figure 7.
Detection of HPV 16 by LCR. HP is control human placental DNA.

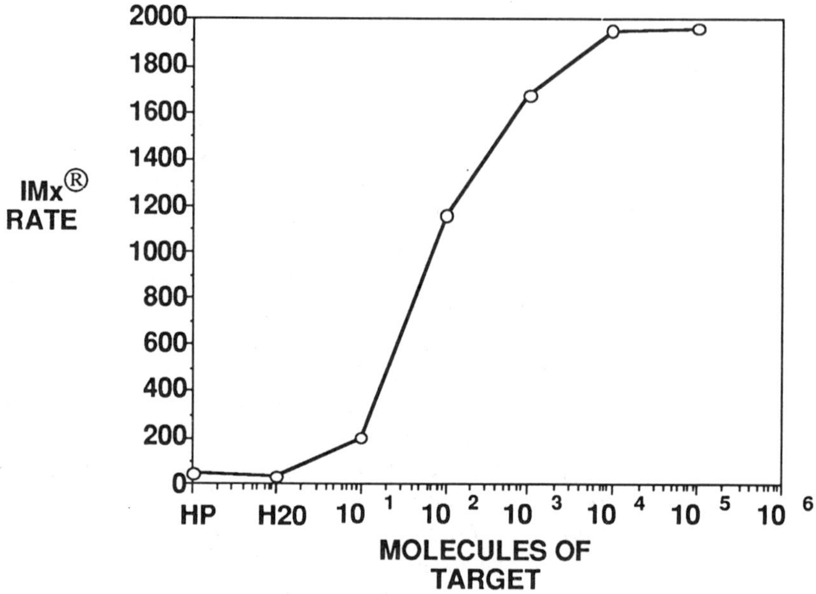

Figure 8.
Detection of HP V 18 by LCR. HP is control human plancental DNA.

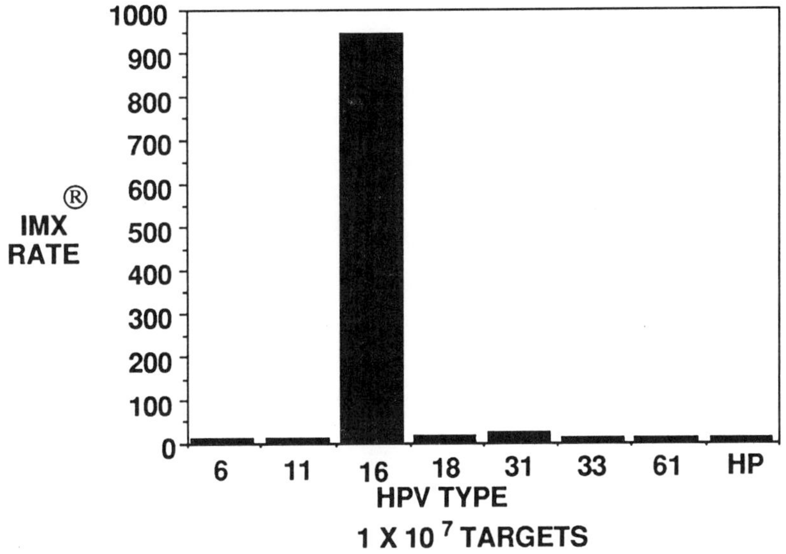

Figure 9.
Assay of various HPV types with probes specific for HPV 16.

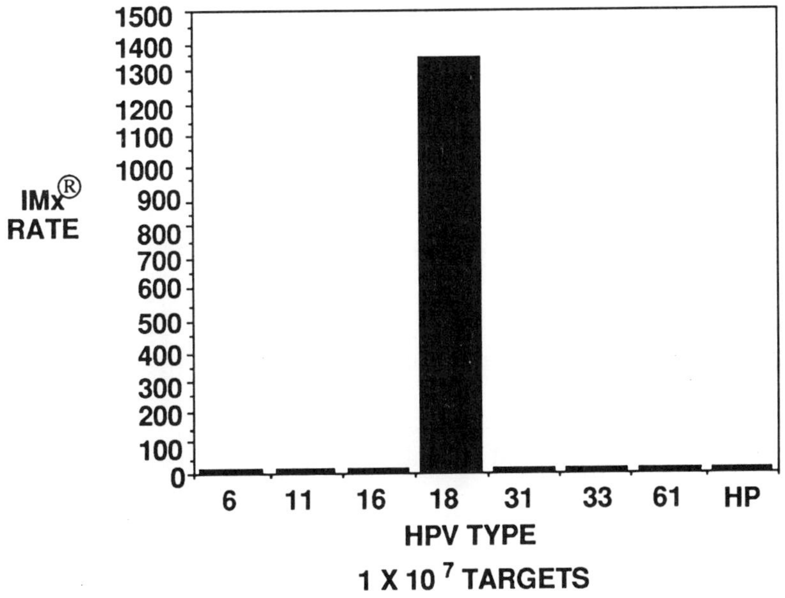

Figure 10.
Assay of various HPV types with probes specific

Sample Type	Southern Blot positive	LCR16 positive	LCR18 positive
HPV16	16	16	0
HPV18	5	0	5
HPV18-like	2	0	0
other HPV	22	0	0
negative control	17	0	0

Figure 11.
Comparison of results of clinical samples typed for HPV 16 and 18 by LCR with typing by Southern Blots. Data from the laboratory of Dr. Wayne Lancaster, Wayne State University, Detroit, Michigan.

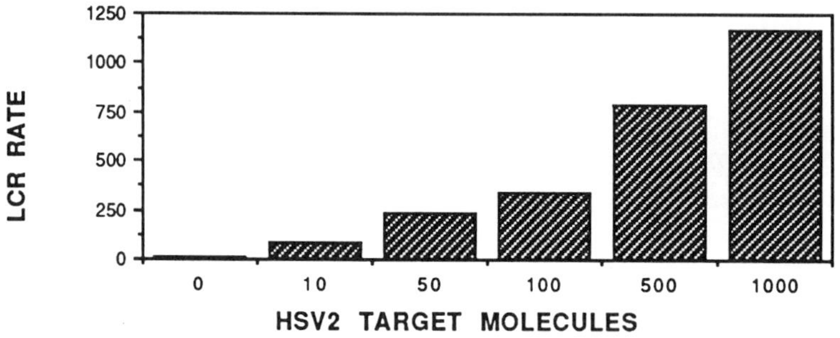

Figure 12.
Detection of HSV–2 plasmid DNA.

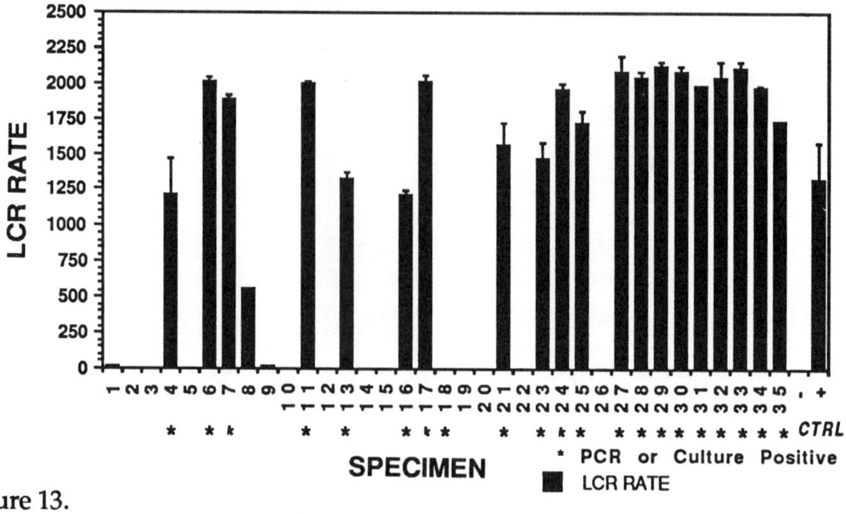

Figure 13.
Detection of HSV–2 in lesions. Good correlation in all but two cases was found with either culture or PCR.

phosphorylated at their 5' ends, i.e., at the point of ligation, with polynucleotide kinase or with the phosphate–ON reagent (Clontech).

The tubes are overlaid with mineral oil and then heated to 100°C for three minutes. A solution containing *T. thermophilus* ligase is then added to bring the final volume to 50 µl. The tubes are then cycled between 85°C and 50°C approximately 30 times (to amplify the DNA) either by alternating between two water baths or using a commercial thermocycler. An 80 µl aliquot of the product is then removed from under the mineral oil and transferred to an IMx^R disposable reaction cell for detection.

The reaction cell is placed in the IMx^R, which has been loaded with sample dilution buffer, anti–fluorescein–coated microparticles, anti–biotin alkaline phosphatase conjugate, 4–methylumbelliferyl phosphate (MUP), and wash solution. On completion of the MEIA protocol, the amount of bound alkaline phosphatase is derived from the reaction rate in counts/sec./sec.

The total assay time is under three hours (Fig. 6).

There is a large body of literature implicating an association between human papilloma virus (HPV) infection and cervical cancer, although a causal relationship has not been established (Koutsky and Wolner–Hanssen, 1989). Certain HPV types (established by DNA hybridization), such as 6 and 11 are associated more commonly with benign lesions, whereas types 16 and 18 have a stronger linkage to malignant progression. For these reasons a sensitive and specific test ameanable to use in the clinical laboratory would have value. Employing LCR technology and the IMx^R immunoassay instrument, as described above, an assay for HPV was developed (Bond *et. al.*, 1989, Perko *et. al.*, 1990). Less than 10 copies of purified HPV plasmids could be detected for either type 16 or 18 (Figs. 7 and 8). The assay could discriminate between types 16 and 18, with no crossreactivity with eight other HPV types (Figs. 9 and 10). The assay also showed good correlation with clinical samples typed by Southern blots (Hampl *et. al.*, 1991, Fig. 11).

Traditional methods for herpes simplex virus (HSV) detection rely on culture or antibody or antigen detection and are either time consuming or lack sensitivity. The prevention of neonatal HSV transmission during delivery and monitoring during antiviral therapy for patents with disseminated HSV infections or encephalitis require more rapid and sensitive procedures. A semi–automated assay employing LCR and the IMx^R is capable of detecting less than 10 HSV target molecules (Rinehardt *et. al.*, 1990) (Fig. 12). The assay can also discriminate between HSV 1 and 2 (data not shown), while consensus probes can detect both types. Good correlation was found for cervical lesions when compared with either cell culture or PCR (Fig. 13).

Due to the potentially large window between infection and seroconversion (Imagawa *et. al.*, 1989) and also low concentration virus in infected blood, a sensitive probe test for HIV has a significant potential application. In particular, the screening of sera from neonates of seropositive mothers has much potential value (Laure *et. al.*, 1988). Figure 14 (Carrino, J., personal communication) illustrates the sensitivity that was obtained with an LCR/IMx^R format for HIV–1 detection; the quanitities of HIV DNA detected correspond to five molecules of a

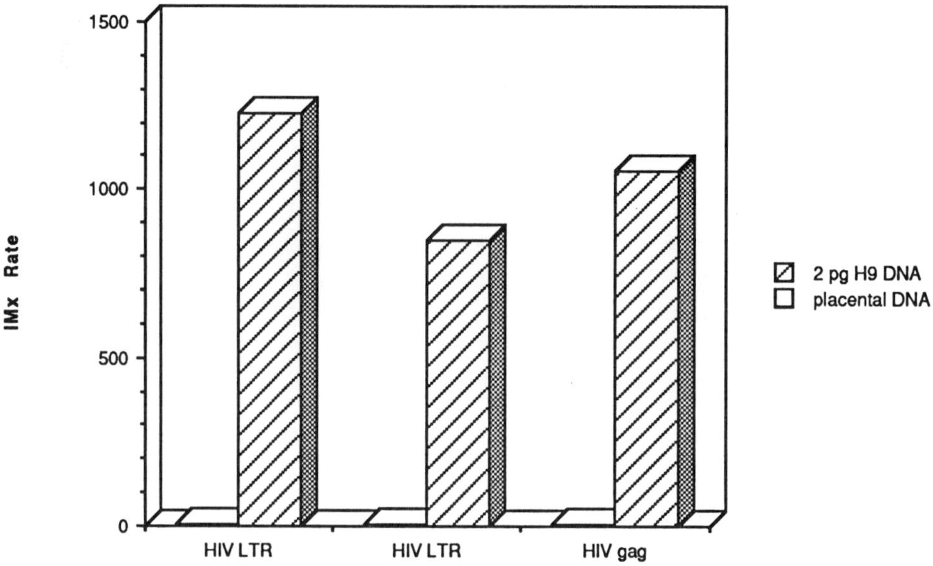

Figure 14.
Detection of different integrated HIV–1 sequences.

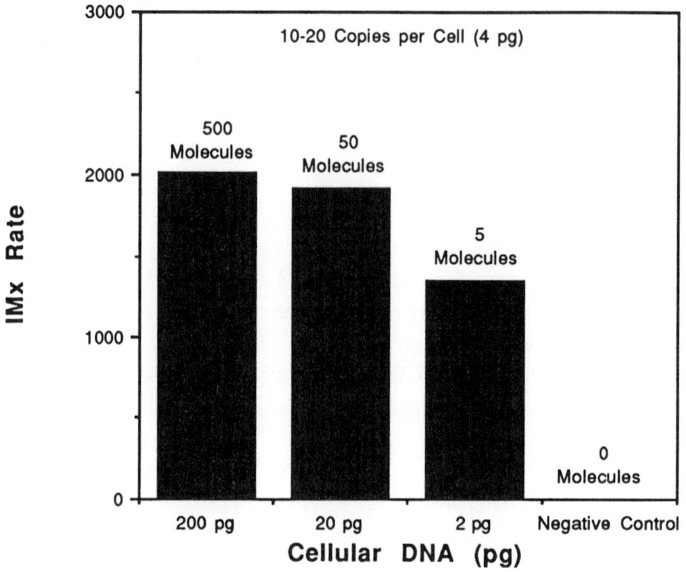

Figure 15.
Detection of integrated HIV–1 DNA from infected H9 cells.

given sequence. The detection of integrated HIV sequences in the H9 lymphoblast cell line is illustrated in Figure 15 (Carrino, J., personal communication), again with five molecule sensitivity.

IV. CONCLUSION

The introduction of DNA amplification technology such as PCR and LCR has allowed the development of nonisotopic probe assays which are rapid, sensitive, specific, and amenable to automation. An example of a semi–automated system was presented here. This system employes LCR technology coupled with a commercial immunoassay system, IMx[R.] Assays have been developed which can detect as little as 10 copies of DNA, with high specificity and be completed in less than three hours. It can be anticipated that as additional automation is developed and the exquisite sensitivity and specificity of these tests is fully appreciated from a clinical standpoint that probe assays will play an increasingly greater role in the clinical laboratory and will make an important contribution to patient care.

REFERENCES

Anonymous, (1991). *J. NIH Res.* **3**, 81–86.

Barany, F. (1991). *Proc. Natl. Acad. Sci. USA* **88**, 189–193.

Bond, S., Carrino, J., Hampl, H., Hanley, K., Rinehardt, L., and Laffler, T. (1990). In *Papillomaviruses in Human Pathology: Recent Progress in Epidermoid Precancers* (J. Monsonego, ed.), pp. 425–434, Raven Press, New York.

Fiore, M., Mitchell, J., Doan, T., Nelson, R. Winter, G., Grandone, C., Zeng, K., Haraden, R., Smith, J., Harris, K., Lexzcynski, J., Berry, D., Safford, S., Barnes, G., Scholnick, A., and Ludington, K. (1988). *Clin. Chem.* **34**, 1726–1732.

Gosling, J. P. (1990). *Clin. Chem.* **36**, 1408–1427.

Hampl, H., Spies, U., Kuhn, J., Dahm, C., Bolik, A., and Rolfs, A. (1991) German Society of Virology, Potsdam.

Imagawa, D. T., Lee, and M. H. Wolinsky, S. M. (1989). *N. Engl. J. Med.* **320**, 1458–1462.

Koutsky, L. A. and Wolner–Hanssen, P. (1989). *Ob. Gynecol. Clinics N. A.* **16**, 541–564.

Kwoh, D. Y. and Kwoh, T. J. (1990). *American Biotechnology Laboratory*, 14–25.

Laure, F., Courgnaud, V., Rouzioux, C., Blanche, S., Veber, F., Burgard, M., Jacomet, C., Griscelli, C., and Brechot, C. (1988). *Lancet* **2**, 538–541.

Landegren, U., Kaiser, R., Caskey, C. T., and Hood, L. (1988). *Science* **242**, 229–237.

Matthews, J. A. and Krika, L. J. (1988). *Anal. Biochem.* **169**, 1–25.

Morgan, R. and Celebuski, J. (1991). *J. Chromatogr.* **536**, 85–93.

Perko, T., Winkler, J., Marshall, R., Jenderzak, D., Campbell, D., and Solomon, N. (1990). *The Fifth San Diego Conference on Nucleic Acids: New Frontiers*.
Rinehardt, L. A., Hampl, H. Carrino, J. J., and Laffler, T. G. (1990). *The Fifth San Diego Conference on Nucleic Acids: New Frontiers.*
Saiki, R. K., Scharf, S., Faloona, F., Mullis, K. B., Horn, G. T., Erlich, H. A., and Arnheim, N. (1985). *Science* **37**, 170–172.
Tenover, F. C. (1988). *Clin. Microbio. Rev.* **1**, 82–101.

11

Current Electron Microscopic Methods in Diagnostic Virology

Robert D. O. Devine

I. INTRODUCTION

The development of electron microscopy as applied to diagnostic virology began with the occasional, yet nonetheless urgent, need to differentiate herpes zoster infections from mild or atypical variola. Laboratory methods had been designed to identify these viruses but only electron microscopy enabled the differentiation to be made within minutes of receiving the clinical specimens. A description of the electron microscopic procedures used to successfully identify herpes zoster or variola in vesicular fluid specimens taken from infected patients was published in 1948 by Nagler and Rake.

The introduction of the negative staining technique using phosphotungstic acid or uranyl acetate (Brenner and Horne 1959, Horne, 1965) provided a rapid and simple method for the elucidation of individual virus particle morphology. Since viruses are identified by their morphologies into herpes–like or adenovirus–like particles etc., the rapid diagnosis of viral infections by electron microscopy (EM) was now possible.

However, the electron microscope of yesteryear was not an instrument which could be manipulated with impunity. Most instruments were mounted on especially constructed plinths, usually in basement stairwells, and each required its own voltage accelerator. Maintenance costs of these early electron microscopes were prohibitive and only a few EM experts could obtain consistent, high quality results. By the late sixties and early seventies, the electronic revolution which brought us the pocket calculator and personal computer had revolutionized electron microscope design. Small, high resolution transmission electron microscopes were developed and made available to diagnostic virology laboratories. It was no longer necessary to spend months mastering the intricacies of column alignment, instrument cleaning and pump

Robert D. O. Devine • Provincial Laboratory of Public Health, University of Alberta, Edmonton, Alberta, Canada T6G 2J2

Applied Virology Research, Vol. 3, Edited by
E. Kurstak *et al.*, Plenum Press, New York, 1994

down sequences. The modern electron microscope was available for immediate use by the non–expert.

Direct visualisation of virus particles in clinical specimens requires an initial virus concentration of 10^6–10^8 particles per ml. This may seem like a prohibitive requirement but it is now recognized that in certain viral infections, viral particles may be found in selected clinical specimens in enormous numbers. The vesicular fluids taken from herpes or poxvirus lesions certainly contain virus in sufficient numbers to permit direct visualisation by electron microscopy and this is also true of fluid stool samples taken from patients with rotavirus gastroenteritis. Indeed, the voluminous body of literature on human gastrointestinal viruses was initiated by the electron microscopic visualisation of human rotavirus (Bishop *et al.* 1973, Flewett *et al.*1973) in the stools of infants suffering from viral gastroenteritis. Since these previously unrecognized viruses did not replicate in commonly used cell cultures, electron microscopy was, until comparatively recently, the only means of laboratory diagnosis. Further electron microscopic investigations of human gastrointestinal disease believed to viral in origin led to the discovery of the enteric adenoviruses (Gary *et al.* 1979), calicivirus (Madely and Cosgrove 1976), astrovirus (Madely and Cosgrove 1975), coronavirus (Caul *et al.*1975) and the numerous small round structured viruses (SRSV) associated with outbreaks of gastrointestinal illness (Appleton *et al.*, 1977; Kapikian *et al.* 1972). It is now recognized that direct electron microscopy in itself may be insufficiently sensitive to detect SRSV in all but the most heavily infected samples (Munroe *et al.*1991) but the use of immunoelectron microscopy (IEM) greatly increases the likelihood of virus detection in stool samples taken from infected patients.

Direct electron microscopic examination of serum samples obtained from patients with serum hepatitis revealed the the Dane particles later shown to be the causative virus of hepatitis B. Recently, aggregates of parvovirus particles (Devine and Leblanc, unpublished) have been visualized in serum taken from patients with Fifth disease. No doubt the search is presently underway for the elusive hepatitis C virus and electron microscopy may well succeed where the traditional methodologies of cell culture have presently failed. However, electron microscopic examination of serum samples is rarely performed nowadays. The sensitive and specific technologies of enzymeimmunoassay or radioimmuno–assay have supplanted the time consuming and labor intensive EM techniques used to detect virus in serum.

There was a number of early publications reporting the suitability of nasopharyngeal secretions for direct EM detection of respiratory syncytial virus, influenzavirus or parainfluenzaviruses, but ordinarily, this author has found that the concentration of virus in these specimens is usually too low to provide consistent results. Attempts to enhance direct EM detection using solid phase immunoelectron microscopy (SPIEM) have proved only partially successful and it is now generally accepted amongst diagnostic virologists that immunofluorescent antibody methods (IFAT) or enzyme immunoassay (EIA) utilising monoclonal antibodies are presently the methods of choice for the rapid detection of respiratory viruses in clinical specimens.

Similarly, direct EM of biopsy specimens , brain liver lung, seldom yields satisfactory results. Rapid demonstration of herpes simplex antigens in brain biopsy tissue taken from patients with herpes encephalitis is much more readily effected by monoclonal IFAT than by direct EM. The same may also be said for lung biopsies taken for cytomegalovirus investigation, although, in this example, inoculation of lung biopsy extract onto shell vials of WI38 fibroblasts and fourteen to eighteen hours incubation is necessary before CMV early antigen can be demonstrated.

Electron microscopic detection of virus particles in all clinical specimens would be ideal, but unfortunately the concentration of viruses in most clinical specimens is far too low. Preliminary procedures for increasing the number or concentration of virus present are therefore often used before the material is examined by electron microscopy. Some clarified specimen extracts may yield positive results after ultracentrifugation (Hammond *et al.*1981) but the majority require inoculation on to susceptible cell monolayers of human or simian origin within which the virus can multiply.

These inoculated cell monolayers are examined daily for the appearance of features indicative of virus infection. When viral replication is suspected, direct electron microscopic examination of the infected monolayer lysate usually reveals the presence and morphology of the infectious agent. Preliminary reports describing the viral agent as resembling herpes, adeno, myxo or other viruses are then issued.

The time saved by applying electron microscopy to cell culture isolation techniques is therefore considerable. Using this combination of techniques it is often possible to make a tentative diagnosis of viral infection and report this is to the patient's physician as early as two days after receipt of the clinical specimen. Thus, in today's diagnostic virology laboratory, electron microscopy retains an eminent role in the rapid detection of viruses either directly from suitable clinical specimens, or after virus replication in permissive cell cultures.

II. ELECTRON MICROSCOPY METHODS

A. Formvar Coating of Copper Grids

1. Fill water trough with dist H_2O until meniscus bulges. Quickly sweep water surface with a glass rod to eliminate surface dust.
2. Take a clean dry microscope slide and dip 1.5 inches vertically into a small beaker containing a 0.30% solution of formvar in ethylene dichloride. Lift vertically and allow to dry.
3. Scratch edges of the formvar coated slide with a scalpel blade.
4. Place the formvar coated slide end at the surface of the water in the prepared water trough. At a constant angle of 45 degrees, push slide down into the water. The formvar film will peel off from the slide and float on the water surface.

5. Place grids copper side up onto the floating formvar film in an orderly fashion (5 rows of 10 grids). Avoid placing grids too closely together and do not place grids too near film edges.

NOTE:
Once a grid has been placed on the formvar film never attempt to rearrange it. Any attempt to do so will result in the loss of all of the grids on the film.
When sufficient grids have been place on the formvar film, remove from the water surface using a clean square of parafilm, and allow to dry. Following formvar coating of the grid, a thin layer of carbon is evaporated onto the formvar to strengthen the grid and provide a measure of contrast.

B. Carbon Coating of Formvar Grids

This procedure will vary according to the vacuum evaporator used for carbon deposition. The following outline serves only as a description of the method used in our laboratory utilising a Varian carbon evaporator.

1. Activate vacuum evaporator and allow to warm up for at least 45 minutes. Prepare carbon rods by finely sharpening one end with the rod sharpener.
2. Tack parafilm backed formvar–coated grids onto a clean dry microscope slide using surface attraction and cut away excess parafilm. Grids face outwards.
3. Place slide on evaporator table with two small pieces of white porcelain upon one of which is a small drop of high vacuum oil used as an indicator of carbon film thickness.
4. Position grid loaded evaporator table for coating by tilting towards the electrodes at an angle of 20 degrees.
5. Align carbon rods on electrodes so that the grids are at the optimum coating distance for the instrument used.
6. Evacuate pressure chamber and evaporate carbon onto the grids using the contrast difference between the oil drop and naked porcelain as a visual guide to carbon film thickness.
7. Repressurize vacuum chamber, remove coated grids and store.

C. Preparation of Specimens for Direct EM

The methods by which specimens are prepared for examination in the electron microscope are varied although in each technique, a single aim is pursued; the electron microscopist must be able to clearly identify individual virus particle morphology.

Specimens suitable for direct EM include vesicle fluids and lesion crusts, fluid stool samples and infected cell culture lysate. Negative staining can be carried out quickly and easily using 1% phosphotungstic acid (PTA) in H_2O, pH

7.0. All EM grid preparation procedures should be carried out within the confines of a Class 1 or greater biohazard hood. Grids are usually prepared in duplicate

1. Vesicle Fluids
   This type of specimen may be applied directly onto a prepared 400 mesh carbon coated formvar grid.
   a. Place a drop of vesicle fluid on the coated side of a prepared grid and blot dry by touching a torn piece of blotting paper to the side of the grid.
   b. Sterilize grid with a drop of 0.25% gluteraldhyde in H_2O and blot dry.
   c. Negatively stain using a drop of 1% PTA, blot dry and examine at 160,000x magnification.

   The negative stain forms an electron dense halo around the electron permissive virus on the grid and highlights typical virus morphology. However, since fluid stool samples or cell culture fluids contain background organic material and salts which tend to crystalize on the EM grids and obscure virus particle morphology, a certain minimum of sample preparation may be necessary. For these specimens there are a number of simple techniques which may be utilized to this end.
2. Fluid Stool Specimens and Not So Fluid Stool Specimens
   a. Fluid stool specimens may be applied directly to a prepared carbon coated grid or after a 1:1 dilution in dd H_2O. Soft stools should be emulsified to liquid form and applied to the grid. Blot dry.
   b. Wash grid by applying a drop of dd H_2O. Blot dry.
   c. Continue 1(b) through 1(c).
3. Biopsy or Tissue Specimens
   Specimens containing large amounts of dissolved salts, i.e. infected biopsy lysate, crust lysate or infected tissue culture cell lysate may be prepared for direct electron microscopy by simple agar dialysis. (Anderson and Doane, 1972)
   a. Place suspect biopsy tissue in a clean sterile metal planchette (a clean 1 oz. universal bottle cap without liner is satisfactory). Add 1 drop of dist H_20 so that tissue is within the fluid.
   b. Freeze and thaw specimen 3 to 5 times by touching planchette alternatively to a block of solid CO_2 and the back of a gloved hand.
   c. Remove one or two drops of specimen lysate taking care not to disturb cellular debris. Place one drop of lysate onto solid 1% Noble agar contained in a microtitre cup.
   d. Place carbon coated formvar grid at the drop surface and allow the fluid to diffuse into the agar. (This removes excessive salts which might crystalize on the grid.)
   e. Remove the grid from the agar surface, sterilize, stain, label and read.

CSF and urine specimens concentrated by ultracentrifugation may also be processed as above.

D. Identification of Virus Particle Morphology

Since the identification of virus particle morphology is not always straightforward, examination of the prepared grids should be carried out by an electron microscopist widely experienced in virus particle morphology.

The classic textbook configuration, e.g. the icosahedral capsomered symmetry of an adenovirus is not always seen in electron microscopic preparations of clinical specimens and diagnosis may have to be made from the observance of non–typical or broken particles. Similarly, herpesvirus virions may be seen intact within an outer membrane or penetrated by the negative stain. Because of their very small size, 25–35 nm, it is often difficult to distinguish complete enterovirus particles from cellular artifacts or ribosomes. However, observance of empty ring–form viral capsids, uniformly 25–35 nm in diameter, is presumptive evidence of enterovirus presence. Other commonly encountered viruses may be identified relatively easily. Thus the double capsid substructure of the rotavirus is easily discerned; the haemagglutinin fringed influenzavirus, unmistakable.

With experience, the electron microscopist can recognize a wide variety of particle morphologies and assign the observed particle to a major virus group.

E. Identification of Enterovirus isolates by Immunoelectron Microscopy

Although the neutralization procedure is still the method of choice to identify unknown enterovirus isolates, the use of immunoelectron microscopy (IEM) to identify enterovirus offers certain advantages during outbreaks of enteroviral disease caused by a single serotype.

In these circumstances a simple immuno–aggregation of the isolated enterovirus may quickly and specifically confirm its identity. The procedure is as follows:

1. Freeze and thaw infected cell cultures.
2. Remove one drop of infected cell lysate and add to one drop of specific neutralizing antisera at a dilution of 20 ND_{50}. Mix well and place two drops of the mixture on to 1% noble agar contained in a microtitre cup.
3. Place carbon coated formvar grid, coated side down on the surface of virus antibody mixture.
4. Incubate the mixture at 37°C for 30 minutes or until the salts etc., have dialysed into the agar.
5. Remove grid, sterilize and negatively stain.

If the enterovirus isolate is aggregated by a specific antiserum e.g.., Echo 14 the virus identity is confirmed. If no virus–antibody aggregates are found and only discrete particles are observed, the isolate must be neutralised in cell culture using the enterovirus typing pools.

In a variation on the above technique viral antisera can be incorporated at optimum dilution into the 1% Noble agar (Anderson and Doane, 1973). This serum in agar methodology combines the agar diffusion procedure with immunoelectron microscopy to give a clearer end result.

F. Solid Phase Immunoelectron Microscopy

Solid phase immunoelectron microscopy (SPIEM) utilizes viral specific antibodies (nowadays mostly monoclonal) to enhance the attachment of virus present in clinical specimens to ionized formvar–carbon coated grids. The antibody solution (optimizes between 1:50 and 1:1000 in water) is coated onto the grid for 10–15 minutes at room temperature. The grids are washed twice in 0.5 M tris buffer pH 7.2 and once with drop of double–distilled H_2O before use. The specimen is incubated with the prepared grid on agar for 20 minutes after which the grid is negatively stained, sterilized and read. A further refinement of this procedure is to first coat the grid using a 1 mg/ml protein A solution prior to coating with antisera (SPIEM–SPA)

Electron dense colloidal gold particles (uniformly 16 nm in diameter) can also be used to signal the presence of specific viral aggregates utilising a technique known as protein A–gold immunoelectron microscopy (PAG– IEM). Protein A solution preparations are electrostatically bonded to the electron dense, easily seen, colloidal gold particles which are then used to signal the presence of antigen–antibody complexes on the EM grid.

G. Infections Diagnosed by Electron Microscopy

Poxviruses

The eradication of smallpox in 1979 forever removed an ancient scourge which had claimed the lives of countless millions down through the ages. Variola major was certainly the most important pathogen of the poxvirus group and the few remaining poxviruses associated with human infection seem trivial by comparison. Nevertheless, diagnostic virology laboratories are often asked to examine clinical specimens for the presence of poxvirus and the electron microscopist must be familiar with the pox and parapoxvirus groups.

Most contemporary requests for poxvirus investigation are associated with zoonotic infections in patients who handle sheep, goats or cattle. Direct EM of vesicle fluid or crusts taken from poxvirus lesions will usually reveal the infectious agent present. Orf is the quintessential parapox virion, 160 x 300nm., spindle rather than brick shaped with surface features reminiscent of a "skein of wool". Vaccinia is an orthopoxvirus, 250 x 350nm and brickshaped.

Other poxviruses of clinical interest include molluscum contagiosum, tanapox and monkeypox. Molluscum contagiosum is a common infection which is usually diagnosed clinically. When laboratory confirmation is requested ,the virus may be visualized by direct EM of an excised lesion. Tanapox virus infections are geographically limited to the Tana river area of Kenya and Zaire. Monkeypox, which may be clinically indistinguishable from smallpox, is due to an orthopox virus found only in the rainforests of Central and West Africa.

Herpesviruses

While there are many infections, systemic and superficial, associated with the human herpesvirus group, (HSV–1 and 2, VZV, CMV, EBV, HHV6) EM diagnosis is essentially limited to the examination of vesicle fluids for typical herpes–like particles. EM diagnosis of herpes simplex or varicella zoster virus infections is not normally given priority in rapid diagnostic laboratories except on those occasions when samples are submitted from neonates or immunocompromised individuals where early diagnosis and chemotherapeutic intervention may be lifesaving.

CMV may be detected in urine samples taken from infected patients after ultracentrifugation and EM of the pellet for typical herpes–like particles. However, it is now recognized that this methodology is much less sensitive than the preferred shell vial technique for CMV early antigen and therefore is rarely used today.

Diarrhoeal Viruses

Over the past seventeen years, it has been well established that type A rotaviruses and, to a lesser extent, adenoviruses are the major causes of gastroenteritis in children under two years (Flewett *et al.*1988; Simon and Mats 1985). In 1987, an epidemic of diarrhoea affecting over two hundred adults in the Chinese city of Qinhuangdao (Fang *et al.* 1989) was found to be due to type B rotavirus also known as adult diarrhoea rotavirus (ADRV).

However, in 20–30% of cases of diarrheal illness in children and a higher proportion of adults with gastrointestinal illness, bacterial or viral agents are seldom detected using standard laboratory procedures (Davidson 1986). Recently, this diagnostic gap has been almost completely closed by the extensive use of electron microscopic methods, (DEM, IEM, SPIEM etc.) and radio or enzyme labeled immunoassays which have demonstrated the presence of one or other of the small round structured viruses (SRSV) Norwalk, calicivirus, astrovirus, etc., in the stools of infected patients. Other viruses including parvo, small round viruses, (SRV) and coronavirus have also been implicated in human diarrheal illness but not to the same extent.

Since virtually all of the SRSV's associated with gastroenteritis do not replicate in commonly used cell cultures, nor is there yet any suitable animal hosts which can provide a model for viral pathogenesis or yield large quantities of virus for research purposes, laboratory diagnosis of these fastidious viruses has been confined to research laboratories or facilities with expertise in electron microscopy. Human volunteers have been the source of the large quantity of virus needed to further elucidate the relationships and characteristics of these important pathogens.

Norwalk-like Agents

These small round structured viruses (SRSV) have been associated with numerous outbreaks of gastroenteritis around the world. (Leers, *et al.*1987, Hayashi *et al.*1989, Kjeldsberg *et al.*1989, Storr, *et al.*1986, Gellert *et al.*1990). Virus transmission is thought to be food or water borne with anal–oral transfer to secondary cases (Sekla *et al.*1989, Riordan *et al.*1984).

Norwalk and associated SRSV's, Hawaii, Snow Mountain and Otofuke agent, were once though to be antigenically distinct but IEM studies using convalescent sera taken from experimentally infected patients have revealed varying degrees of antigenic crossrelatedness (Madore, *et al.*1990). SPIEM tests, using whole serum and purified IgM obtained from volunteers experimentally infected with a Norwalk–like strain which had been recovered from infected patients in geographically separate outbreaks of diarrheal illness in the United Kingdom, indicated that there are at least three Norwalk–like serotypes (Lewis, 1990). A report from Japan, suggests that there may be as many as nine serotypes with varying degrees of cross–reactivity to the candidate Norwalk and Otofuke strains (Okada *et al.* 1990).

Virus obtained from human volunteers experimentally infected with Norwalk virus was used to construct recombinant complementary DNA (cDNA) and clones representing most of the viral genome were obtained. The specificity of the clones was shown by their hybridization with virus obtained from symptomatic volunteers, and with purified Norwalk virus. Hybridisation assays between overlapping clones, restriction enzyme analyses and partial nucleotide sequencing all indicate that Norwalk virus contains a single stranded, 7.5 kilobase RNA genome of positive sense with a polyadenylated tail at the 3' end. A consensus amino acid sequence typical of RNA–dependent RNA polymerase was also identified (Xi *et al.*1990).

Caliciviruses

Calicivirus are small round structured particles (35–40 nm) with a distinctive appearance by electron microscopy. They are fastidious viruses which do not replicate in existing cell culture systems. Caliciviridae have been associated with diarrheal illness in humans, farm animals and reptiles. In vivo experiments confirm that the virus replicates in the gut and is species specific. Molecular analysis indicates caliciviruses are RNA viruses possessing a single major structural poly peptide of 60–71 kD (Cubitt, 1987).

Astroviruses

Astroviruses are so named for their distinctive star like morphology on electron microscopy. They have been associated with outbreaks of gastroenteritis across the globe. (Kurtz *et al.*1977, Ashley *et al.*1978, Konno *et al.*1982). There appears to be at least five astrovirus serotypes, all of which have been cultivated after blind passage in trypsin treated cell cultures (Lee and Kurtz 1981). A monoclonal antibody EIA which detects the presence of all known astrovirus types with a specificity and sensitivity of 96% and 91% respectively has been developed (Herrmann *et al.*1990) and used to ascertain the role of astrovirus infections in daycare facilities (Lew *et al.*1991). It seems likely therefore that the laboratory diagnosis of astrovirus infections will rely less on EM and more on EIA in future studies of astrovirus associated illness.

Coronaviruses

Coronaviruses were first thought to be only associated with mild upper respiratory infections (Tyrrell and Bynoe 1965, Hamre and Procknow 1966) Later observations of corona –like particles in stool samples taken from patients with gastroenteritis have aroused speculation that coronavirus is a major cause of human diarrheal illness.

The human enteric coronaviruses are between 80 and 160nm in diameter with distinctive 20nm club shaped projections extending across the surface of the particle. They have not yet been isolated in cell cultures.

REFERENCES

Anderson, N., Doane, F.W. (1972). *Appl. Microbiol.* **24,** 95–496.
Anderson, N., Doane, F.W.(1973). *Can.J. Microbiol.* **19,** 585–589.
Appleton , H (1987). *CIBA Found. Symp.* **128,** 108–25.
Appleton, H., Buckley, M., Thom, B.T. Cotton, J.L. and Henderson, S.(1977). *Lancet* **1,** 409.
Bishop, R.F., Davidson, G.P., Holmes, H. (1973). *Lancet* **2,** 1497.
Brenner, S. and Horne, R.W. (1959). *Biochim. Biophys. Acta.* **34,**103–110.
Caul, E.O., Paver, W.K. and Clark, S.K.R.(1975). *Lancet* **1,** 1192.
Caul, E.O., Ashley, C.R. and Egglestone, S.I. (1977). *Med. Lab. Sci.* **34,** 259–263.
Cubitt WD (1987). *CIBA Found. Symp.* **128,** 26–43.
Cubitt WD (1990). *Clin. Gastroenterol.* **4,** 643–56.
Fang, Z.Y., Ye, Q. HD, M.S., Dong, H., Qing, S., Penaranda, M.E., Wen, I., Glass, R.T., (1989). *J. Infect. Dis.* **160,** 948–953.
Flewett, T.H., Bryden, A.S. and Davies, H. (1973). *Lancet* **2,** 1497.
Flewett, T.H., Beards, G.M., Brown,D.W., Sanders, R.C.(1987). *CIBA Found. Symp.* **128,** 238–249.
Gary, G.W., JNR., Hierholzer, J.C. and Black, R.E. (1979). *J.Clin. Microbiol.* **10,** 96.
Gordon, S.M., Oshiro, L.S., Jarvis. W.R., Donenfeld, D., Ho, M.S., Taylor, F., Greenberg, H.B., Glass, R., Madore, H.P., Dolin, R., (1990). *Am. J.Epidemiol.* **131,** 702–710.
Hammond, G.W., Hazelton, P.R., Chuang, I., Klisko, B. (1981). *J. Clin. Microbiol.* **14,** 210–221.
Hayashi, Y., Ando, T., Utagawa, E., Sekine, S., Okada, S., Yabuuchi, K., Miki, T., Ohash, M. (1989). *J. Clin. Microbiol.* **27,** 1728–1733.
Herrmann., J.E.,(1984). *Lancet,* **2,** 1405.
Herrmann, J.E., Nowak, N.A., Perron–Henry, D.M., Hudson, R.W., Cubitt, W.D., Blacklow, N.R. (1990). *J. Infect. Dis.* **161,** 226–229.

Hopley, J.F.A., Doane, F.W. (1985). *J. Virol. Meth.* **12,** 135–147.

Horne, R.W., (1965). In *Techniques for Electron Microscopy* (R. Bullock and P. Petrusz, eds.) pp. 328–355. Blackwell Scientific, Oxford.

Kapikian, A.Z., Wyatt, R.G., Dolin, R., Thornhill, T.S., Kalica, A.R., and Chanock, R.M., (1972). *J. Virol.* **10,** 1075.

Katz, D., Straussman, Y., Shahar, A., Kohn, A., (1980). *J. Immunol. Meth.* **38,** 171–174.

Kjeldsberg, E. (1986). *J. Virol. Meth.* **14,** 321–333.

Kjeldsberg, E., Anestad, G., Greenberg, H., Orstavik, I., Pedersen, R., Sletteb,E. (1989). *Scand. J. Infect. Dis.* **21,** 521–526.

Konno, T., Suzuki, H., Ishida, N., Chiba, R., Mochizuki, K., Tsunoda, A., (1982). *J. Med. Virol.* **9,** 11–17.

Lee, T.W., Kurtz, J.B., (1981). *J. Gen. Virol.* **57,** 421–424.

Leers, W.D., Kasupski, G., Fralick, R., Wartman, S., Garcia, J., Gary, W., (1987). *Am J. Public Health.* **77,** 291–295.

Lew, J.F., Moe, C.L., Munroe, S.S., Allen, J.R., Harrison, B.M., Forrester, B.D., Stine, S.E., Woods, P.A., Hierholzer, J.C., Herrmann, J.E., Blacklow, N.R., Bartlett, A.V., Glass, R.I., (1991). *J, Infect. Dis.***164,** 673–678.

Lewis, D.C., Lightfoot, N.F., Pether, J.V., (1988). *J. Clin. Microbiol.* **26,** 938–942.

Lewis, D.C., (1990). *J. Med. Virol.* **30,** 77–81.

Madely, C.R. and Cosgrove, P.B. (1976). *Lancet* **1,** 199.

Madore, H.P., Treanor, J.J., Buja, R., Dolin, R (1990). *J. Med. Virol.* **32,** 96–101.

Monroe, S.S., Glass, R.I., Noah, N., Flewett, T.H., Caul, E.O., Ashton, C.I., Curry, A., Field, A.M., Madely, R., Pead, P.J.(1991). *J. Med. Virol.* **33,** 193–198.

Madore, H.P., Treanor, J.J., Buja, R., Dolin, R (1990). *J. Med. Virol.* **32,** 96–101.

Nagler, F.P.O. and Rake, G (1948). *J. Bacteriol.* **55,** 45–51.

Okada, S., Sekine, S., Ando, T., Hayashi, Y., Murao, M., Yabuuchi, K., Miki, T., Ohashi, M., (1990). *J. Clin.. Microbiol.* **28,** 1244–1248.

Polak, J.M., Varndell, I.M. eds. (1984). *Immunolabelling for Electron Microscopy.* Elsevier, New York .

Roth, J., (1983). In *Techniques in Immunochemistry* Vol. 2, 215. ??????

Sekla, L., Stackiw, W., Dzogan, S., Sargeant, D. (1989). *Can. Med. Assoc. J.* **149,** 1461–1464.

Storr, J., Rice, S., Phillips, A.D., Price, E., Walker–Smith, J.A., (1986). *J. Pediatr. Gastroenterol. Nutr.* **5,** 576–580.

Index